ALGAE BASED BIOELECTROCHEMICAL SYSTEMS FOR CARBON SEQUESTRATION, CARBON STORAGE, BIOREMEDIATION, AND BIOPRODUCT GENERATION

Bioelectrochemical Systems:
The Way Forward – Volume III

ALGAE BASED BIOELECTROCHEMICAL SYSTEMS FOR CARBON SEQUESTRATION, CARBON STORAGE, BIOREMEDIATION, AND BIOPRODUCT GENERATION

Edited by

DURGA MADHAB MAHAPATRA
Department of Chemical Engineering, Energy Cluster, School of Advanced Engineering (SoAE), University of Petroleum and Energy Studies (UPES), Dehradun, Uttarakhand, India

SMITA S. KUMAR
Department of Environmental Sciences, J.C. Bose University of Science & Technology, YMCA, Faridabad, Haryana, India

Department of Environmental Studies, Central University of Haryana, Jant-Pali, Mahendergarh, Haryana, India

LAKHVEER SINGH
Associate Professor, Department of Chemistry, Sardar Patel University, Mandi, Himachal Pradesh, India

ELSEVIER

ACADEMIC PRESS
An imprint of Elsevier

Academic Press is an imprint of Elsevier
125 London Wall, London EC2Y 5AS, United Kingdom
525 B Street, Suite 1650, San Diego, CA 92101, United States
50 Hampshire Street, 5th Floor, Cambridge, MA 02139, United States

Notices
Knowledge and best practice in this field are constantly changing. As new research and experience broaden our understanding, changes in research methods, professional practices, or medical treatment may become necessary.

Practitioners and researchers must always rely on their own experience and knowledge in evaluating and using any information, methods, compounds, or experiments described herein. In using such information or methods they should be mindful of their own safety and the safety of others, including parties for whom they have a professional responsibility.

To the fullest extent of the law, neither the Publisher nor the authors, contributors, or editors, assume any liability for any injury and/or damage to persons or property as a matter of products liability, negligence or otherwise, or from any use or operation of any methods, products, instructions, or ideas contained in the material herein.

ISBN 978-0-323-91023-1

For information on all Academic Press publications
visit our website at https://www.elsevier.com/books-and-journals

Publisher: Charlotte Cockle
Acquisitions Editor: Peter Adamson
Editorial Project Manager: Mason Malloy and Franchezca A. Cabural
Production Project Manager: Fizza Fathima
Cover Designer: Greg Harris

Typeset by STRAIVE, India

Contents

Contributors

Fazal Adnan
Atta Ur Rahman School of Applied Biosciences, National University of Sciences & Technology, Pakistan

Aarti Ahlawat
Department of Chemistry, PDM University, Bahadurgarh, Haryana, India

Junaid Ahmad
Department of Microbiology, Hazara University, Mansehra, Pakistan

Gagan Anand
Applied Science Cluster, SOAE, University of Petroleum and Energy Studies, Dehradun, India

Sandhya Babel
School of Bio-Chemical Engineering and Technology, Sirindhorn International Institute of Technology, Thammasat University, Pathum Thani, Thailand

Kulvinder Bajwa
University School for Graduates Studies, USGS, Chaudhary Devi Lal University, Sirsa, India

Muhammad Yousuf Jat Baloch
College of New Energy and Environment, Jilin University, Changchun, China

Rachna Bhateria
Department of Environmental Sciences, Maharshi Dayanand University, Rohtak, Haryana, India

Bhavna
Advanced Sensors Lab, Department of Physics, Deenbandhu Chhotu Ram University of Science and Technology, Sonipat, Haryana, India

Narsi R. Bishnoi
Department of Environmental Science and Engineering, Guru Jambheshwar University of Science and Technology, Hisar, India

Aryan Boora
Advanced Sensors Lab, Department of Physics, Deenbandhu Chhotu Ram University of Science and Technology, Sonipat, Haryana, India

Bansal Deepak
JBM Group, Gurugram, Haryana, India

Anuj Rana
Department of Microbiology, College of Basic Sciences & Humanities, CCS Haryana Agricultural University, Hisar, India

Vineeta Rana
Department of Botany and Plant Physiology (Environmental Science), College of Basic Sciences & Humanities, CCS Haryana Agricultural University, Hisar, India

Annu Rani
Department of Microbiology, College of Basic Sciences & Humanities, CCS Haryana Agricultural University, Hisar, India

Dolly Rani
Department of Microbiology, College of Basic Sciences & Humanities, CCS Haryana Agricultural University, Hisar, India

Kajal Saini
Department of Environmental Sciences, J.C. Bose University of Science & Technology, YMCA, Faridabad, Haryana, India

Supriya Sehrawat
Advanced Sensors Lab, Department of Physics, Deenbandhu Chhotu Ram University of Science and Technology, Sonipat, Haryana, India

Muhammad Ajmal Shah
Department of Pharmacy, Hazara University, Mansehra, Pakistan

Jyoti Sharma
Central Pollution Control Board, New Delhi, India

Preeti Sharma
University Institute of Engineering and Technology, Maharshi Dayanand University, Rohtak, India

Sudheesh K. Shukla
School of Biomedical Engineering, Shobhit Institute of Engineering and Technology, Meerut, India; Department of Chemical Sciences, University of Johannesburg, Doornfontein Campus, Johannesburg, South Africa

Swayam Siddha
School of Environment and Sustainable Development, Central University of Gujarat, Gandhinagar, Gujrat, India

Muhammad Faisal Siddiqui
Department of Microbiology, Hazara University, Mansehra, Pakistan

Lakhveer Singh
Department of Chemistry, Sardar Patel University, Mandi, Himachal Pradesh, India

Rajesh Singh
Environmental Hydrology Division, National Institute of Hydrology, Roorkee, India

Sandeep Singh
Environmental Hydrology Division, National Institute of Hydrology, Roorkee, India

Manaswee Suttipong
Department of Chemical Technology, Faculty of Science, Chulalongkorn University, Bangkok, Thailand

Ihsan Ullah
Department of Biological Sciences, Faculty of Science, King Abdulaziz University, Jeddah, Saudi Arabia

Shefali Upadhyay
Department of Environmental Sciences, J.C. Bose University of Science & Technology, YMCA, Faridabad, Haryana, India

Anjali Yadav
Department of Environmental Science and Engineering, Guru Jambheshwar University of Science & Technology, Hisar, Haryana, India

Chamath D.Y. Yahampath Arachchige Don
School of Bio-Chemical Engineering and Technology, Sirindhorn International Institute of Technology, Thammasat University, Pathum Thani, Thailand

CHAPTER 1

Integration of bioelectrochemical and algal systems for bioproducts generation

Chamath D.Y. Yahampath Arachchige Don and Sandhya Babel
School of Bio-Chemical Engineering and Technology, Sirindhorn International Institute of Technology, Thammasat University, Pathum Thani, Thailand

1 Introduction

Bioelectrochemical systems (BESs) employ microorganisms or enzymes as biocatalysts with electrochemical methods to generate electricity or other useful products such as hydrogen, methanol, ethanol, hydrogen peroxide, and formic acid [1–3]. Microbial fuel cell (MFC) is a widely used bioelectrochemical device that can convert chemical energy stored in bonds of organic substrates into electricity using electrochemically active microorganisms [4,5]. The biogas production from wastewater followed by biogas conversion to electricity has significant energy loss (>60%), resulting in an inefficient overall process [6]. Thus, MFC provides a promising method for the direct conversion of the energy stored in organic compounds to electricity.

On the other hand, the cultivation of microalgae to recover nutrients in algal biomass is a prominent biological wastewater treatment method that can be accomplished in open and closed systems [7,8]. Photoautotrophic microalgae manufacture organic carbon from inorganic carbon when light is available as the energy source. The overall photosynthesis reaction can be written as Eq. (1). Integration of the algal with bioelectrochemical systems can enhance the bioproduct generation from wastewater when combining the features attributing to both technologies as discussed in the following sections.

$$6CO_2 + 6H_2O + \text{Photons (light energy)} \rightarrow \text{biomass} + 6O_2 \qquad (1)$$

2 Bioproduct generation of two-chamber microbial fuel cell utilizing algal cathode chambers

In two-chamber MFC systems where oxygen is used as the electron acceptor in the cathode chamber, the required dissolved oxygen in the catholyte is generally provided through mechanical aeration. Mechanical aeration is an energy-intensive process that reduces the sustainability of the system. As a result, the photosynthetic oxygen generation

Algae Based Bioelectrochemical Systems for Carbon Sequestration, Carbon Storage, Bioremediation and Bioproduct Generation
https://doi.org/10.1016/B978-0-323-91023-1.00005-3

in the cathode chamber appears as an alternative to the mechanical supply of oxygen. In this case, the algal-assisted cathode chamber generates oxygen through the photosynthesis processes of microalgae [9–13]. In this system, the electrochemically active bacteria attached to the anode oxidize the organic compounds to generate electrons and protons as in a typical MFC system. Possible oxidation reactions for acetate, glucose, and glycerol are shown in Eqs. (2)–(4) [14], respectively. The protons travel through the separator membrane from the anodic chamber to the cathodic chamber, while electrons travel through an external circuit to produce current. In the cathodic chamber, microalgae in the presence of light and an inorganic carbon source such as carbon dioxide produce oxygen via photosynthesis. On the cathode, a reaction occurs between electrons, protons, and oxygen to form water as shown in Eq. (5) [15]. The working principles of an algal-assisted MFC are shown in Fig. 1.

$$CH_3COO^- + 3H_2O \longrightarrow CO_2 + HCO_3^- + 8H^+ + 8e^- \tag{2}$$

$$C_6H_{12}O_6 + 6H_2O \longrightarrow 6CO_2 + 24H^+ + 24e^- \tag{3}$$

$$C_3H_8O_3 + 6H_2O \longrightarrow 3HCO_3^- + 17H^+ + 14e^- \tag{4}$$

$$O_2 + 4H^+ + 4e^- \longrightarrow 2H_2O \tag{5}$$

In a typical MFC, electricity is the bioproduct that generates through the anodic bioelectrochemical processes. However, integrating algal-assisted cathode chamber has the additional benefit of generating algal biomass as a bioproduct simultaneously. Regardless

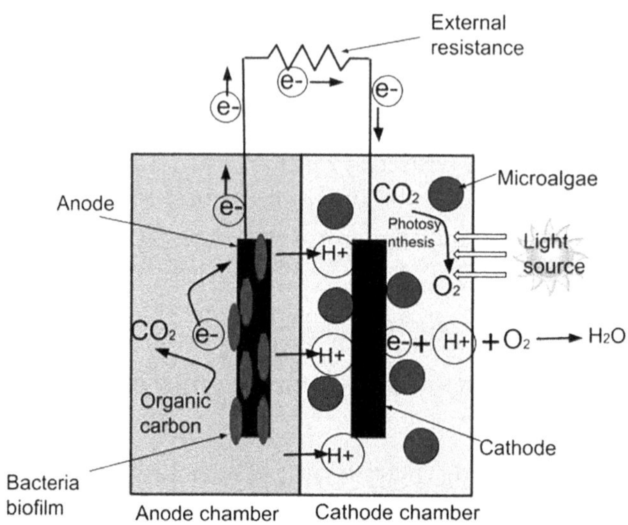

Fig. 1 Two-chamber microbial fuel cell integrating algal cathode chamber generating oxygen via photosynthesis.

of the different methods of oxygen supply process in the cathode chamber, the final electron acceptor is oxygen in both mechanical aeration and photosynthetic aeration systems. The algal cathode chambers have been shown better performance due to the pure oxygen generation via photosynthesis rather than aerating atmospheric air, which contains 21% of oxygen approximately. An algal-assisted cathode chamber generated 34.5% higher power density and the internal resistance of the system was decreased by 13.7% when integrating an algal cathode chamber compared to the mechanical aeration [16]. In another study, the dissolved oxygen (DO) concentration in the catholyte increased to $15.7\,mg\,L^{-1}$ under the photosynthetic aeration with *Scenedesmus obliquus*, resulting in a maximum power density of $153\,mW\,m^{-2}$. The DO and maximum power density were decreased by 63% and 24% in mechanical aeration, respectively [17].

Different two-chamber MFC models have been investigated in the literature. The most common architecture is the rectangular-shaped MFC with two identical chambers as shown in Fig. 1. A reactor consisting of two identical chambers generated a power density of $3720\,mW\,cm^{-3}$ with *Chlorella vulgaris* assisting the carbon fiber cloth modified with platinum ($0.1\,mg/cm^2$) cathode [18]. Using a tubular cathodic chamber in an MFC, a maximum power density of $24.4\,mW\,m^{-2}$ was obtained with *C. vulgaris* assisting the carbon paper-coated, platinum cathode [19].

Generally, the cathode chamber is provided with a feed containing nutrients and the anode chamber is provided with a feed containing organic carbon. Some studies attempted to circulate the effluent of the anode chamber to the cathode chamber as the growth medium for microalgae (Fig. 2). Ma et al. [20] developed a photosynthetic MFC with a stainless-steel mesh, which serves as both the cathode and the filtration device. In this system, anode effluent is directed to an algae cathode chamber for biomass

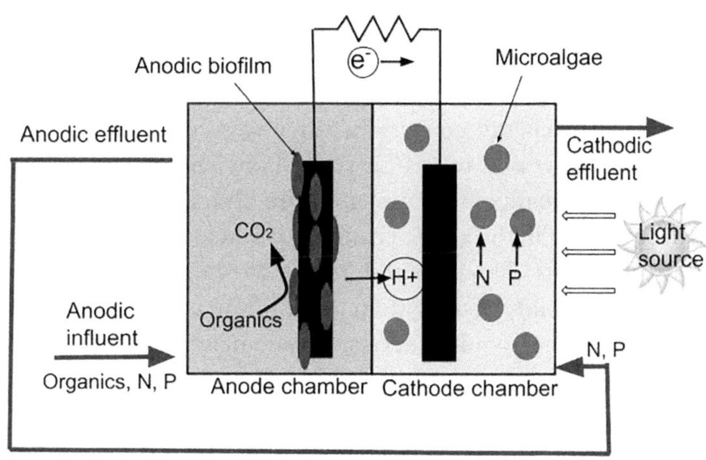

Fig. 2 Two-chamber microbial fuel cell for the subsequent treatment of anodic effluent in the algal cathode chamber.

generation, and biomass yield rates of 32.7 and 60.8 $mg L^{-1} d^{-1}$ were achieved with flow rates of 0.3 and 0.6 $mL min^{-1}$, respectively. The maximum current density reached approximately 200 $mA m^{-2}$. In an MFC assisted by *Chlorella vulgaris* in which the 10-day-treated anode effluent is used as the catholyte, a maximum power density of 9.956 $mW m^{-2}$ and biomass productivity of 0.374 $g L^{-1} d^{-1}$ were obtained treating wastewater containing NH_4^+-N and COD concentrations of 314 and 1500 $mg L^{-1}$ [21]. These findings show that the anodic effluent, which is rich in nutrients, is suitable to grow microalgae in the cathode chamber. However, appropriate hydraulic retention times (HRTs) should be provided for the anodic bacteria to reduce the organic carbon contents from the feed to ensure photoautotrophic growth of microalgae rather than heterotrophic growth which limits the oxygen generation.

2.1 Bioelectricity generation

2.1.1 Effects of the anodic chamber on bioelectricity generation

Bioelectricity generation in the cell depends on the oxidation reaction on the anode and the reduction reaction on the cathode. Anodic reaction occurred by microorganisms oxidizes the organic carbon to carbon dioxide. The main factors related to the anode chamber that influences the current generation in an MFC are the types of the bacterial community in the anodic biofilm [22,23] and the concentration/loading rate of the electron donor (organic compound) in the anodic chamber [24]. These factors have not been well investigated with algal-incorporated MFCs. However, since the anodic conditions are similar in algal-incorporated MFCs and conventional MFCs, the anodic influences on the bioelectricity generation can be characterized with in-depth literature available on MFC studies.

The microbial inoculum type in the anode chamber is one of the important parameters in MFC applications. Several bacteria species such as *Shewanella oneidensis MR-1* [25], *Geobacter sulfurreducens* [26], *Aeromonas hydrophila* [27], and *Rhodoferax ferrireducens* [28] have demonstrated transfer of the electron to the anode using different mechanisms. *G. sulfurreducens* resulted in a low internal resistance of 165 Ω, high Coulombic efficiency (CE) of 25.8%, and high energy conversion efficiency (ECE) of 7.2%. However, the internal resistance was increased to 620 Ω, and CE and ECE were decreased to 9.2% and 2.4%, respectively, with mixed culture inoculum [29]. In algal-assisted MFCs, mixed cultures are mainly utilized in the anode chamber collected from the biological sludge of wastewater treatment plants. The high cost associated with operating pure cultures can reduce the economic feasibility of the system. In addition, mixed culture inoculums have the advantage of consuming a wide range of organic matters which is beneficial in real applications of treating wastewater.

2.1.2 Effects of the algal cathode chamber on bioelectricity generation

In algal-assisted MFCs, the method of the oxygen supply for the cathodic reduction reaction is through the photosynthesis of microalgae. DO concentration in the catholyte directly affects the bioelectricity generation of the cell. A higher DO of

$12.29 \pm 1.47\,mg\,L^{-1}$ was observed in the catholyte with the *Chlorella vulgaris*, where the growth of algae was stimulated by adding $100\,mg\,L^{-1}$ of organic carbon (COD) in addition to other essential nutrients such as nitrogen and phosphorus. In contrast, the DO was slightly lower ($10.70 \pm 2.15\,mg\,L^{-1}$), when no organics were added to the catholyte. As a result, the maximum power density increased from $5.91 \pm 0.50\,mW\,m^{-2}$ (no organics) to $9.07 \pm 1.40\,mW\,m^{-2}$ ($100\,mg\,L^{-1}$ of COD) [30]. Table 1 summarizes the bioelectricity

Table 1 Maximum power densities and catholyte DO concentrations were obtained in microbial fuel cells with an algal-assisted cathode chamber.

Cathode material	Microalgae inoculum	Cathode chamber volume (mL)	MPD[a]	DO ($mg\,L^{-1}$)	References
Carbon cloth	*Scenedesmus acutus*	75	1.3[b]	20–25	[31]
Pt-based gas diffusion electrode	*S. acutus*	75	5.3[b]	20–25	[31]
Graphite plate	*Chlorella vulgaris*	100	34.2[b] (383.6[c])	24	[11]
Pt-coated carbon paper	*Scenedesmus obliquus*	250	153[b]	5.9	[17]
Carbon fiber brush	Mixed culture	255	50[b] (517[c])	19.57	[32]
Graphite	*Synechococcus* sp.	300	95.63[b]	8.6	[33]
Carbon paper coated with $0.6\,mg/cm^2$ Pt catalyst	*C. vulgaris*	300	24.4[b]	20.8	[19]
Carbon felt	*C. vulgaris*	300	54.48[b]	11.2 ± 0.24	[34]
Graphite felt	*C. vulgaris*	650	21[b]	3.2–3.7	[35]
Graphite	Mixed culture	797	3.35[b]	4.5	[36]
Carbon cloth with 10% of Teflon	*C. vulgaris*	800	42.98[b]	7	[12]
Plain carbon cloth	Mixed culture	800	11.5[b]	6–8	[37]
Graphite plate	*C. vulgaris*	1000	29.55[b]	10–11	[16]
Carbon fiber cloth	*C. vulgaris*	1000	7.13[b]	10.70 ± 2.15	[30]
Carbon fiber cloth	*Chlorella pyrenoidosa*	1000	268[c]	13.6–14.8	[38]

[a]Maximum power density.
[b]Power normalized to the surface area of the cathode electrode ($mW\,m^{-2}$).
[c]Power normalized to the volume of the cathode chamber ($mW\,m^{-3}$).

generations achieved in algal MFC systems and the corresponding DO in the catholyte. Evidently, *C. vulgaris* has been studied in algal MFCs more than any other species.

Initial biomass concentration has an influence on electricity generation. Maximum power densities of 455.6 ± 21.4 and $466.9 \pm 24.4\,\mathrm{mW\,m^{-3}}$ were obtained at initial microalgal biomass of 0.50 and $0.75\,\mathrm{g\,L^{-1}}$, respectively. These were 7.2% and 9.9% higher than a lower initial microalgal biomass of $0.25\,\mathrm{g\,L^{-1}}$ in the catholyte [39]. Technologically, improved cathode materials can enhance the catalytic properties of biocathodes. Traditionally, platinum is used due to its high oxygen reduction reaction (ORR) catalytic activity. A platinum-based electrode generated a higher maximum power density of $5.3\,\mathrm{mW\,m^{-2}}$ compared to the carbon cloth electrode ($1.3\,\mathrm{mW\,m^{-2}}$ utilizing *Scenedesmus acutusas* in the cathode chamber [31]). However, considering the cost of platinum, other types of composite materials are widely investigated attempting to reach the same magnitude of power generation as platinum. A bare nickel modified with graphene was employed as a cathode by Zhang et al. [40] for *Chlorella* sp. assisted MFC. It was observed that the power density increased by eightfold compared to the bare nickel.

Light intensity and light regimes are some of the major factors that influence the growth of microalgae and their composition including fatty acid and pigment profiles [41]. In algal MFCs, alternative light and dark regimes are used to simulate natural solar light conditions. During the light conditions, due to the photosynthesis of algae, the DO in the catholyte reached high, resulting in high electricity generation [12]. However, no oxygen was added to the catholyte by microalgae during the dark conditions as a result of ceased photosynthesis. The remaining DO in the catholyte is used by microalgae for aerobic respiration, resulting in a rapid drop in the electricity output during dark cycles. Bazdar et al. [9] compared different light/dark regimes of 24/00, 16/8, and 12/12 h, and observed the maximum power densities of 126, 112, and $72\,\mathrm{mW\,m^{-3}}$, respectively. Compared to the intermittent illuminations of 16/8 and 12/12 h, the maximum power density was 12.7% and 74.8% higher under continuous illumination.

At an illumination intensity of 10,000 lx, a higher DO in the catholyte was observed compared to a lower intensity of 5000 lx and a higher intensity of 15,000 lx [39]. In another study, an intensity of 5000 lx resulted in the highest power density of $126\,\mathrm{mW\,m^{-3}}$. The power density dropped when increasing the light intensity to 10,000 lx, and decreasing to 3500 lx [9]. Lower or higher illumination intensities can affect the growth of the microalgae cells due to photolimitation and photoinhibition, respectively [42]. A PMFC was operated using *C. vulgaris* in the cathode chamber by Gouveia et al. [43] with two different light intensities. Increasing the light intensity from 26 to $96\,\mathrm{\mu E\,m^{-2}\,s^{-1}}$ led to an increase of about sixfold in the power produced.

The microalgae are sensitive to light and can affect the catholyte DO concentration. Catholyte DO is an important parameter, which directly affects the power outputs of the cell. Thus, the optimum light intensity for the photosynthetic growth of the selected microalgae culture should be provided to enhance the performance. Even though more stable power generation can be obtained under continuous light conditions, light/dark regimes are more suitable for practical applications considering the availability of sunlight.

2.2 Factors influencing algal biomass generation as a bioproduct

The wastewater contains various organic carbons, nitrogen, and phosphorus in different forms. The ability of microalgae to grow using a certain compound varies depending on the species. Therefore, a suitable species should be selected according to the characteristic of the cathodic feed. Table 2 provides a comparison of biomass generation of different microalgae species under different conditions in the cathode chamber.

Light intensity and the light/dark regime can significantly influence biomass generation in the cathode chamber. A higher biomass growth rate was observed for *Synechococcus* sp. than *Chlorococcum* sp. under 800 lx of light intensity using a BG11 nutrient medium that contained nitrogen in the form of nitrate [46]. A higher biomass productivity of $0.58\,g\,L^{-1}\,d^{-1}$ was obtained under continuous light conditions compared to light/dark regimes of 12/12 h ($0.51\,g\,L^{-1}\,d^{-1}$) and 32/32 h ($0.39\,g\,L^{-1}\,d^{-1}$ [16]). Higher biomass concentrations can be obtained under continuous light due to continuous biomass generation. However, in practical applications, alternative light/dark regimes are suitable where sunlight can provide energy for photosynthesis rather than providing artificial light to avoid energy consumption.

Mixotrophic cultivation of microalgae can result in higher biomass generation rather than autotrophic cultivation due to its faster growth rate [47]. In the mixotrophic mode, microalgae simultaneously use inorganic carbon and organic carbon for autotrophic photosynthesis and heterotrophic aerobic respiration, respectively [48,49]. The biomass productivity of $0.104\,g\,L^{-1}\,d^{-1}$ obtained in the autotrophic mode increased to $0.176\,g\,L^{-1}\,d^{-1}$ with $300\,mg\,L^{-1}$ of COD concentration in the catholyte with *C. vulgaris* [30]. However, higher organic carbon concentration can limit the photosynthetic oxygen generation of microalgae due to the enhanced heterotrophic metabolisms which can lead to decreased electricity generation.

It was also demonstrated that temperature variation can affect microalgae growth significantly. A higher biomass concentration in the cathode chamber was obtained during spring ($4.2\,g\,L^{-1}$) compared to summer ($0.9\,g\,L^{-1}$) [50]. The seasonal variations are problematic in long-term field applications unlike the laboratory studies conducted in controlled environments. Therefore, the biomass generation of the system under natural conditions in the long term should be investigated further in future studies.

3 Coupling the MFC with an external algal reactor

The cathodic chamber of an MFC can be coupled with an external photobioreactor (PBR) to enhance the DO in the catholyte as shown in Fig. 3A. Under 14/10 h of light/dark regimes, the light phase voltage reached more than 0.5 V, and it dropped

Table 2 Biomass generation of different microalgae species under different conditions in the cathode chamber.

Characteristics of the cathodic feed	Algae inoculum	C-HRT[a]	Light conditions	BP[b] ($g L^{-1} d^{-1}$)	References
BG11 medium without Na_2CO_3 and budding 10% (v/v) CO_2 mixed with 90% N_2.	*Chlorella vulgaris*	17	Continuous light, 3500 lx of intensity	0.0996	[44]
Nutrient medium containing 300 $mg L^{-1}$ of KNO_3 and 0.25 $g L^{-1}$ of initial algal biomass	*C. vulgaris*	16.7	Continuous light, 7000 lx of intensity	0.58	[16]
Synthetic wastewater containing NH_4^+-N of 300 $mg L^{-1}$ and initial algal biomass of 0.25 $g L^{-1}$	*C. vulgaris*	19	Continuous light, 7000 lx of intensity	0.104	[30]
Anaerobically digested swine wastewater (COD = 486–662 $mg L^{-1}$, TN[c] = 705–788 $mg L^{-1}$)	*Chlorella pyrenoidosa*	8–10	Continuous light, 5000 lx of intensity	0.59	[38]
50% municipal solid waste leachate (COD = 1283.7 $mg L^{-1}$, TN[c] = 86.45 $mg L^{-1}$)	*Synechococcus* sp.	16	Continuous light	0.1275	[33]
Nutrient medium containing 100 $mg L^{-1}$ of $NaNO_3$	*C. vulgaris*	11	12/12 h of dark/light regimes, 106.6 $\mu mol m^{-2} s^{-1}$ of light intensity	0.1066	[34]
Nutrient medium containing 300 $mg L^{-1}$ of KNO_3 and 0.25 $g L^{-1}$ of initial algal biomass	*C. vulgaris*	16	12/12 h of dark/light regimes, 7000 lx	0.51	[16]
Effluent wastewater from a chocolate factory containing COD of 7002 $mg L^{-1}$ and initial algal biomass of 3.76 $g L^{-1}$	*C. vulgaris*	32	Solar radiation ($1100 \pm 225 W m^{-2}$)	0.1625	[45]
The synthetic wastewater containing 188 mg of $C_6H_{12}O_6$ and 134 mg of NH_4Cl was pumped into the anode chamber and the anodic effluent circulated to the cathode chamber at a flow rate of 0.55 $mL min^{-1}$, initial algal biomass of 0.75 $g L^{-1}$	*C. vulgaris*	11.8	Continuous light, 10,000 lx of intensity	0.3757	[39]
Synthetic wastewater (NH_4^+-N = 314 $mg L^{-1}$, COD = 1500 $mg L^{-1}$) treated in the anodic chamber circulated to the cathode chamber in batch mode, initial algal biomass of 0.25 $g L^{-1}$	*C. vulgaris*	10	Continuous light, 7000 lx of intensity	0.374	[21]

[a]C-HRT = Hydraulic retention time in the cathode chamber.
[b]BP = Biomass productivity.
[c]TN = Total nitrogen.

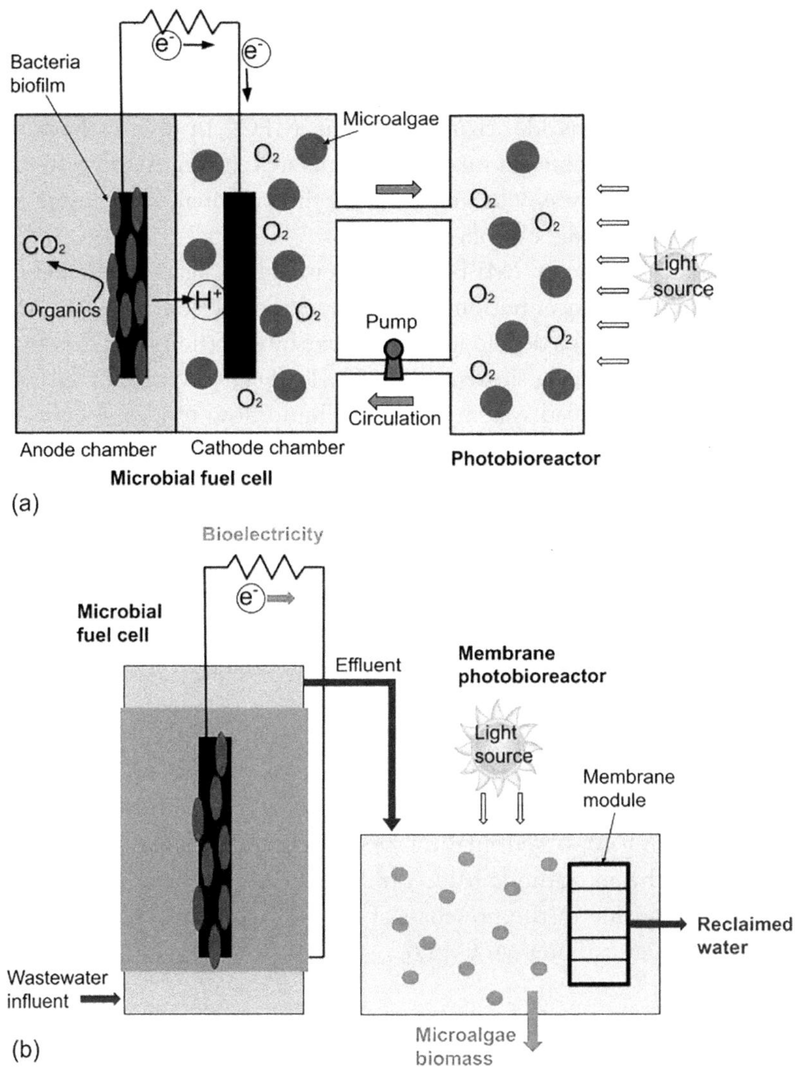

Fig. 3 (A) Two-chamber microbial fuel cell with the cathode chamber coupled with a photobioreactor. (B) A tubular microbial fuel cell and membrane photobioreactor coupled system.

to 0.15 V during the dark phase. When recirculating the catholyte through a PBR, the difference in the light and dark voltage output was minimized by reaching the dark phase voltage of more than 0.45 V providing a more stable system [51].

Jiang et al. [52] developed a coupled system consisting of an up-flow membrane-less MFC and a PBR. In this system, a glass wool layer and glass bead layer were used as the separator between the two chambers. The cathodic effluent was continuously pumped into a column PBR that contained mixed culture microalgae. A maximum volumetric power density of 481 mW m^{-3} was achieved with the combined removal of 99.3% of total phosphorus (TP) and 99.0% NH_4^+-N in the coupled system. Nayak and Ghosh [53] have used the pretreated wastewater in a PBR (utilizing *Scenedesmus abundans*) for further treatment in the anode chamber of an MFC. In the cathode chamber, *S. abundans* was grown in a nutrient medium to generate oxygen. In this system, a power density of 838.68 mW m^{-2} was achieved showing the potential of using the effluent of PBR as the feed for an anodic chamber.

Membrane photobioreactor (MPBR) is an emerging algae-based technology to obtain higher biomass concentrations and productivities [54,55]. An MFC- and MPBR-coupled system can be developed to generate bioelectricity and harvest algal biomass while reclaiming the waster shown in Fig. 3B. Tse et al. [56] conducted a study with an MFC- and MPBR-coupled system with two liquid flow modes, where in the first system, the effluent from the anode chamber flowed directly into the MPBR and there was a separate liquid stream acting as a catholyte. In the second system, anodic effluent flowed to the cathode chamber and functioned as the catholyte, before being supplied to the MPBR as the algal growth medium. The algal biomass production in the second system was decreased due to the conversion of ammonium to nitrate during the nitrification process on the cathode. Ammonium is considered the most convenient nitrogen source for cell metabolism; thus, the microalgae preferred ammonium over other forms of nitrogen such as nitrate [57,58].

The photosynthetic DO generated in a PBR can be collected and provided for the air cathode of single-chamber MFCs as the electron acceptor as shown in Fig. 4. A coupled system of an air–cathode MFC and an algae bioreactor showed a maximum power density of 0.63 W m^{-2} using 40% O_2 from the algae bioreactor and it was approximately 30% higher compared to the air–cathode MFC utilizing the atmospheric air. The increase in power generation was due to the presence of high oxygen concentration in the algae reactor compared to air cathode MFC [59].

4 Two-chamber microbial fuel cell using algae in the anode chamber

Algae have been used in the anode chamber as a substrate to provide organic carbon for anodic bacteria [43,60,61]. *Chlorella regularis* biomass that contained easily biodegradable proteins (46%) and carbohydrates (22%) in the anode chamber generated a power density of 0.86 W m^{-2} and Columbic efficiency of 61.5% [62]. However, these microalgae are dead algal biomass and acted as the primary electron donor as any other form of an organic compound.

In some studies, live microalgae are employed in the anode chamber as the electron donor. However, oxygen in the anode chamber can directly accept the electrons without transferring them through the external circuit to the cathode. Therefore, oxygen generation by microalgae in the anode chamber should be minimized. It was demonstrated that

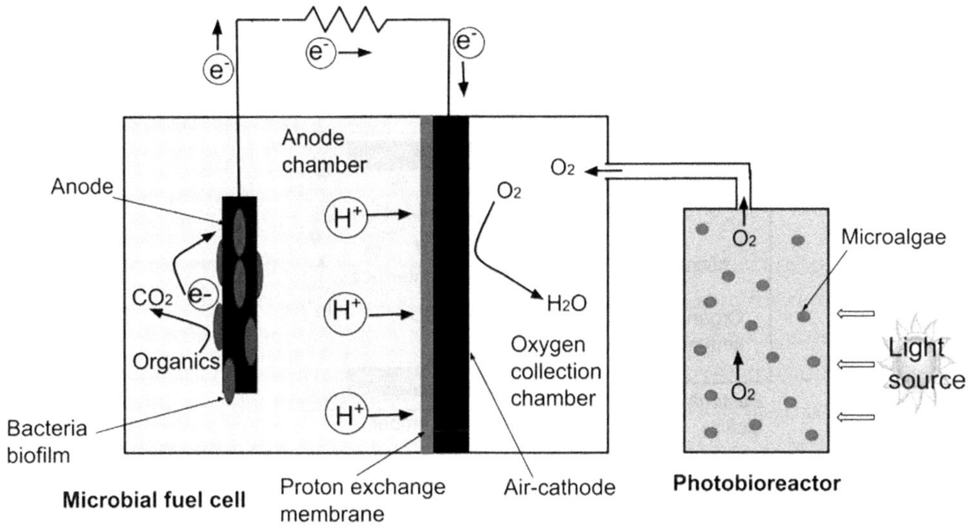

Fig. 4 A single-chamber air–cathode microbial fuel cell coupled with an algae bioreactor to generate oxygen for the cathodic reduction reaction.

lower light intensities can significantly enhance power generation due to the low DO production. With live microalgae at the anode as an electron donor and potassium ferricyanide at the cathode as the electron acceptor, the maximum power density of the cell reached $6030\,\mathrm{mW\,m^{-2}}$ under a light intensity of 2500 lx. Power generation was decreased with a high light intensity of 6500 lx due to an increase in the DO from $2.7\,\mathrm{mg\,L^{-1}}$ (under a light intensity of 2500 lx) to $4.9\,\mathrm{mg\,L^{-1}}$. It was demonstrated that electricity output can be enhanced significantly by adding sodium sulfite to the anolyte to absorb oxygen [63].

Instead of common proton exchange membranes, the anode and cathode chambers were separated by a gas-permeable membrane by Longtin et al. [64]. In this system, oxygen (electron acceptor) is produced in the anode chamber by photosynthesis of microalgae. Oxygen then diffuses through the gas-permeable membrane to the cathode where the reduction reaction occurs. Connecting three cells in parallel, a maximum power density of $59.8 \pm 7.9\,\mathrm{mW\,cm^{-2}}$ was obtained in a staked MFC system. However, due to the availability of oxygen in the anode chamber, some electrons can be diminished in the anolyte before being harvested as electricity.

5 Sediment microbial fuel cell with algae assisting the cathode

In sediment MFC (SMFC), the anode is placed in organic-rich anaerobic sediment and the cathode is placed in the overlying aerobic water [65]. There is no membrane separator as in typical two-chamber MFCs; however, unlike single-chamber MFC that utilizes air

respectively, 0.307 $kg\,m^{-3}\,d^{-1}$ algal productivity, 0.09 $kg\,m^{-3}\,d^{-1}$ lipid productivity, and 11.5318 $kWh\,m^{-3}$ of net energy recovery (NER) were achieved [74].

The commercialization of power generation and integration with algal systems is not economically feasible due to some limitations in MFC technology [75]. The power output generated from an individual MFC unit is insufficient for practical applications. Stacked MFC systems connecting individual cells in series or parallel are widely investigated to increase the power output in practical applications while maintaining the same magnitude of power generation as laboratory-scale MFC systems. However, voltage reversal and ionic short-circuiting are some of the major drawbacks in stacked systems in series [76]. In the parallel connection, the voltage output cannot be enhanced significantly as in the series connection even though the system provides fewer complications [77]. In this regard, Yang et al. [78] developed a stacked algal-assisted MFC system consisting of multiple anodic chambers installed in an algal raceway pond with the assistance of capacitors. The maximum power density reached 2.34 $W\,m^{-3}$ with capacitors, which was 77% higher than that without capacitors proving a promising approach to practical applications of MFC-coupled algal systems.

In practical applications due to the seasonal temperature variations, the power output of the system can be varied. An MFC with mixed culture inoculated cathode chamber showed a higher power generation of 57.0 $Mw\,m^{-2}$ during the spring. It was drastically decreased to 1.1 $mW\,m^{-2}$ during the summer [50]. In addition, when the algal systems were conducted under solar radiation, the voltage significantly dropped during the night time due to aerobic respiration. Reaching the stability of the power generation is a major challenge in MFC utilizing an algal aerated cathode chamber compared to mechanical aeration. It was demonstrated that partially submerged biocathodes can enhance the power generation in the cell. Significantly higher voltages were obtained in an algae-based MFC system when only 50% of the electrode is submerged rather than completely submerged electrodes. The partially submerged electrode can obtain the atmospheric oxygen when the DO generated by photosynthesis is not sufficient for the cathodic reduction reaction proving a more stable power generation [38].

8 Conclusions

Integrating an algae-based cathode chamber to generate oxygen for the cathodic reduction reaction has been demonstrated with enhanced bioelectricity generation. In addition, algal biomass can be harvested as a feedstock for different products such as fuels, power, materials, and chemicals. Light intensity and temperature affect the biomass productivity of the system. Providing continuous light for the microalgae resulted in a stable electricity output and higher biomass productivities. However, to reduce energy consumption, sunlight can be employed with alternative light and dark regimes. However, the major disadvantage of light and dark regimes is the drop of the DO in the cathode chamber, leading to a drop in the electricity output. Coupling an external PBR provides a promising prospect to the MFC technology with enhanced stability of the cell. The DO accumulated through the photosynthesis of microalgae in PBRs during the daytime can be circulated to the cathode chamber of MFCs during the nighttime to compensate for the reduced DO in the catholyte. Live algae have been investigated in the anode chamber as the electron donor. However, the accumulation of photosynthetic DO in the anodic chamber should be minimized to avoid electron acceptance by oxygen. Sediment microbial fuel cells and photosynthetic microbial desalination cells are other bioelectrochemical systems that have been investigated for coupling with an algal cathode chamber for better performance. However, in practical applications, the various environmental conditions in long-term operation can lead to system complication due to the microalgae's sensitivity to light and temperature. Pilot studies are needed to address these technical issues before scaling up for real-world applications.

References

[1] S. Bajracharya, M. Sharma, G. Mohanakrishna, X. Dominguez Benneton, D.P.B.T.B. Strik, P.M. Sarma, D. Pant, An overview on emerging bioelectrochemical systems (BESs): technology for sustainable electricity, waste remediation, resource recovery, chemical production and beyond, Renew. Energy 98 (2016) 153–170.

[2] D. Pant, A. Singh, G. Van Bogaert, S. Irving Olsen, P. Singh Nigam, L. Diels, K. Vanbroekhoven, Bioelectrochemical systems (BES) for sustainable energy production and product recovery from organic wastes and industrial wastewaters, RSC Adv. 2 (4) (2012) 1248–1263.

[3] T. Zheng, J. Li, Y. Ji, W. Zhang, Y. Fang, F. Xin, W. Dong, P. Wei, J. Ma, M. Jiang, Progress and prospects of bioelectrochemical systems: electron transfer and its applications in the microbial metabolism, Front. Bioeng. Biotechnol. 8 (2020) 10.

[4] B.E. Logan, Exoelectrogenic bacteria that power microbial fuel cells, Nat. Rev. Microbiol. 7 (5) (2009) 375–381.

[5] B.E. Logan, B. Hamelers, R. Rozendal, U. Schröder, J. Keller, S. Freguia, P. Aelterman, W. Verstraete, K. Rabaey, Microbial fuel cells: methodology and technology, Environ. Sci. Technol. 40 (17) (2006) 5181–5192.

[6] B.E. Rittmann, Opportunities for renewable bioenergy using microorganisms, Biotechnol. Bioeng. 100 (2) (2008) 203–212.

[7] F. Ji, Y. Liu, R. Hao, G. Li, Y. Zhou, R. Dong, Biomass production and nutrients removal by a new microalgae strain Desmodesmus sp. in anaerobic digestion wastewater, Bioresour. Technol. 161 (2014) 200–207.

[8] E.-S. Salama, M.B. Kurade, R.A.I. Abou-Shanab, M.M. El-Dalatony, I.-S. Yang, B. Min, B.-H. Jeon, Recent progress in microalgal biomass production coupled with wastewater treatment for biofuel generation, Renew. Sust. Energ. Rev. 79 (2017) 1189–1211.

[9] E. Bazdar, R. Roshandel, S. Yaghmaei, M.M. Mardanpour, The effect of different light intensities and light/dark regimes on the performance of photosynthetic microalgae microbial fuel cell, Bioresour. Technol. 261 (2018) 350–360.

[10] J.M. Burkholder, P.M. Glibert, H.M. Skelton, Mixotrophy, a major mode of nutrition for harmful algal species in eutrophic waters, Harmful Algae 8 (1) (2008) 77–93.

[11] A.S. Commault, O. Laczka, N. Siboni, B. Tamburic, J.R. Crosswell, J.R. Seymour, P.J. Ralph, Electricity and biomass production in a bacteria-Chlorella based microbial fuel cell treating wastewater, J. Power Sources 356 (2017) 299–309.

[12] A. Gonzalez del Campo, J.F. Perez, P. Cañizares, M.A. Rodrigo, F.J. Fernandez, J. Lobato, Characterization of light/dark cycle and long-term performance test in a photosynthetic microbial fuel cell, Fuel 140 (2015) 209–216.
[13] M.K. Enamala, R. Dixit, A. Tangellapally, M. Singh, S.M.P. Dinakarrao, M. Chavali, S.R. Pamanji, V. Ashokkumar, A. Kadier, K. Chandrasekhar, Photosynthetic microorganisms (algae) mediated bioelectricity generation in microbial fuel cell: concise review, Environ. Technol. Innov. 19 (2020) 100959.
[14] E. Oguz Koroglu, H. Civelek Yoruklu, A. Demir, B. Ozkaya, Chapter 3.9 - scale-up and commercialization issues of the MFCs: challenges and implications, in: S.V. Mohan, S. Varjani, A. Pandey (Eds.), Microbial Electrochemical Technology, Elsevier, Amsterdam, 2019, pp. 565–583.
[15] Z. Du, H. Li, T. Gu, A state of the art review on microbial fuel cells: a promising technology for wastewater treatment and bioenergy, Biotechnol. Adv. 25 (5) (2007) 464–482.
[16] C.D.Y.Y.A. Don, S. Babel, Comparing the performance of microbial fuel cell with mechanical aeration and photosynthetic aeration in the cathode chamber, Int. J. Hydrog. Energy 46 (31) (2020) 16751–16761.
[17] R. Kakarla, B. Min, Photoautotrophic microalgae Scenedesmus obliquus attached on a cathode as oxygen producers for microbial fuel cell (MFC) operation, Int. J. Hydrog. Energy 39 (19) (2014) 10275–10283.
[18] Y. Zhang, Y. Zhao, M. Zhou, A photosynthetic algal microbial fuel cell for treating swine wastewater, Environ. Sci. Pollut. Res. 26 (6) (2019) 6182–6190.
[19] X.Y. Wu, T.S. Song, X.J. Zhu, P. Wei, C.C. Zhou, Construction and operation of microbial fuel cell with Chlorella vulgaris biocathode for electricity generation, Biotechnol. Appl. Biochem. 171 (8) (2013) 2082–2092.
[20] J. Ma, Z. Wang, J. Zhang, T.D. Waite, Z. Wu, Cost-effective Chlorella biomass production from dilute wastewater using a novel photosynthetic microbial fuel cell (PMFC), Water Res. 108 (2017) 356–364.
[21] C.D.Y.Y.A. Don, S. Babel, Circulation of anodic effluent to the cathode chamber for subsequent treatment of wastewater in photosynthetic microbial fuel cell with generation of bioelectricity and algal biomass, Chemosphere 278 (2021) 130455.
[22] K.L. Lesnik, H. Liu, Establishing a core microbiome in acetate-fed microbial fuel cells, Appl. Microbiol. Biotechnol. 98 (9) (2014) 4187–4196.
[23] Y. Qu, Y. Feng, X. Wang, B.E. Logan, Use of a coculture to enable current production by geobacter sulfurreducens, Appl. Environ. Microbiol. 78 (9) (2012) 3484.
[24] A. Fraiwan, D.F. Call, S. Choi, Bacterial growth and respiration in laminar flow microbial fuel cells, J. Renew. Sustain. Energy 6 (2) (2014) 023125.
[25] Y.A. Gorby, S. Yanina, J.S. McLean, K.M. Rosso, D. Moyles, A. Dohnalkova, T.J. Beveridge, I.S. Chang, B.H. Kim, K.S. Kim, Electrically conductive bacterial nanowires produced by Shewanella oneidensis strain MR-1 and other microorganisms, Proc. Natl. Acad. Sci. U.S.A. 103 (2006) 11358.
[26] G. Reguera, K.D. McCarthy, T. Mehta, J.S. Nicoll, M.T. Tuominen, D.R. Lovley, Extracellular electron transfer via microbial nanowires, Nature 435 (2005) 1098.
[27] C.A. Pham, S.J. Jung, N.T. Phung, J. Lee, I.S. Chang, B.H. Kim, H. Yi, J. Chun, A novel electrochemically active and Fe(III)-reducing bacterium phylogenetically related to Aeromonas hydrophila, isolated from a microbial fuel cell, FEMS Microbiol. Lett. 223 (2003) 129.
[28] S.K. Chaudhuri, D.R. Lovley, Electricity generation by direct oxidation of glucose in mediatorless microbial fuel cells, Nat. Biotechnol. 21 (10) (2003) 1229–1232.
[29] D. Jiang, B. Li, W. Jia, Y. Lei, Effect of inoculum types on bacterial adhesion and power production in microbial fuel cells, Biotechnol. Appl. Biochem. 160 (1) (2010) 182–196.
[30] C.D.Y.Y.A. Don, S. Babel, Effects of organic loading on bioelectricity and micro-algal biomass production in microbial fuel cells using synthetic wastewater, J. Water Process. Eng. 39 (2021) 101699.
[31] S. Angioni, L. Millia, P. Mustarelli, E. Doria, M.E. Temporiti, B. Mannucci, F. Corana, E. Quartarone, Photosynthetic microbial fuel cell with polybenzimidazole membrane: synergy between bacteria and algae for wastewater removal and biorefinery, Heliyon 4 (3) (2018) e00560.

[32] H.T.H. Nguyen, R. Kakarla, B. Min, Algae cathode microbial fuel cells for electricity generation and nutrient removal from landfill leachate wastewater, Int. J. Hydrog. Energy 42 (49) (2017) 29433–29442.

[33] R. Lakshmidevi, N.N. Gandhi, K. Muthukumar, Carbon neutral electricity production from municipal solid waste landfill leachate using algal-assisted microbial fuel cell, Biotechnol. Appl. Biochem. 191 (2) (2020) 852–866.

[34] G. Yadav, I. Sharma, M. Ghangrekar, R. Sen, A live bio-cathode to enhance power output steered by bacteria-microalgae synergistic metabolism in microbial fuel cell, J. Power Sources 449 (2020) 227560.

[35] D.-B. Wang, T.-S. Song, T. Guo, Q. Zeng, J. Xie, Electricity generation from sediment microbial fuel cells with algae-assisted cathodes, Int. J. Hydrog. Energy 39 (25) (2014) 13224–13230.

[36] D.F. Juang, C.H. Lee, S.C. Hsueh, Comparison of electrogenic capabilities of microbial fuel cell with different light power on algae grown cathode, Bioresour. Technol. 123 (2012) 23–29.

[37] J. Lobato, A. González del Campo, F.J. Fernández, P. Cañizares, M.A. Rodrigo, Lagooning microbial fuel cells: a first approach by coupling electricity-producing microorganisms and algae, Appl. Energy 110 (2013) 220–226.

[38] J. Ling, Y. Xu, C. Lu, W. Lai, G. Xie, L. Zheng, M.P. Talawar, Q. Du, G. Li, Enhancing stability of microalgae biocathode by a partially submerged carbon cloth electrode for bioenergy production from wastewater, Energies 12 (17) (2019).

[39] Y. Wang, Z. Lin, X. Su, P. Zhao, J. Zhou, Q. He, H. Ai, Cost-effective domestic wastewater treatment and bioenergy recovery in an immobilized microalgal-based photoautotrophic microbial fuel cell (PMFC), Chem. Eng. J. 372 (2019) 956–965.

[40] Y. Zhang, Q. HeL, Y. Xia, S.S. Li, Algae cathode microbial fuel cells for cadmium removal with simultaneous electricity production using nickel foam/graphene electrode, Biochem. Eng. J. 138 (2018) 179–187.

[41] Z. Amini Khoeyi, J. Seyfabadi, Z. Ramezanpour, Effect of light intensity and photoperiod on biomass and fatty acid composition of the microalgae, Chlorella vulgaris, Aquac. Int. 20 (1) (2012) 41–49.

[42] B. Kokabian, V.G. Gude, Sustainable photosynthetic biocathode in microbial desalination cells, Chem. Eng. J. 262 (2015) 958–965.

[43] L. Gouveia, C. Neves, D. Sebastião, B.P. Nobre, C.T. Matos, Effect of light on the production of bioelectricity and added-value microalgae biomass in a photosynthetic alga microbial fuel cell, Bioresour. Technol. 154 (2014) 171–177.

[44] M. Li, M. Zhou, J. Luo, C. Tan, X. Tian, P. Su, T. Gu, Carbon dioxide sequestration accompanied by bioenergy generation using a bubbling-type photosynthetic algae microbial fuel cell, Bioresour. Technol. 280 (2019) 95–103.

[45] R. Huarachi-Olivera, A. Dueñas-Gonza, U. Yapo-Pari, P. Vega, M. Romero-Ugarte, J. Tapia, L. Molina, A. Lazarte-Rivera, D.G. Pacheco-Salazar, M. Esparza, Bioelectrogenesis with microbial fuel cells (MFCs) using the microalga Chlorella vulgaris and bacterial communities, Electron. J. Biotechnol. 31 (2018) 34–43.

[46] S. Naina Mohamed, P. Ajit Hiraman, K. Muthukumar, T. Jayabalan, Bioelectricity production from kitchen wastewater using microbial fuel cell with photosynthetic algal cathode, Bioresour. Technol. 295 (2020) 122226.

[47] S. Kim, J.E. Park, Y.B. Cho, S.J. Hwang, Growth rate, organic carbon and nutrient removal rates of Chlorella sorokiniana in autotrophic, heterotrophic and mixotrophic conditions, Bioresour. Technol. 144 (2013) 8–13.

[48] O. Perez-Garcia, F.M.E. Escalante, L.E. De-Bashan, Y. Bashan, Heterotrophic cultures of microalgae: metabolism and potential products, Water Res. 45 (1) (2011) 11–36.

[49] J. Zhan, J. Rong, Q. Wang, Mixotrophic cultivation, a preferable microalgae cultivation mode for biomass/bioenergy production, and bioremediation, advances and prospect, Int. J. Hydrog. Energy 42 (12) (2017) 8505–8517.

[50] S. Venkata Mohan, S. Srikanth, P. Chiranjeevi, S. Arora, R. Chandra, Algal biocathode for in situ terminal electron acceptor (TEA) production: synergetic association of bacteria–microalgae metabolism for the functioning of biofuel cell, Bioresour. Technol. 166 (2014) 566–574.

[51] I. Gajda, A. Stinchcombe, J. Greenman, C. Melhuish, I. Ieropoulos, Algal 'lagoon' effect for oxygenating MFC cathodes, Int. J. Hydrog. Energy 39 (36) (2014) 21857–21863.
[52] H.-M. Jiang, S.-J. Luo, X.-S. Shi, M. Dai, R.-B. Guo, A system combining microbial fuel cell with photobioreactor for continuous domestic wastewater treatment and bioelectricity generation, J. Cent. South Univ. 20 (2) (2013) 488–494.
[53] J.K. Nayak, U.K. Ghosh, Post treatment of microalgae treated pharmaceutical wastewater in photosynthetic microbial fuel cell (PMFC) and biodiesel production, Biomass Bioenergy 131 (2019) 105415.
[54] M.R. Bilad, D. Vandamme, I. Foubert, K. Muylaert, I.F.J. Vankelecom, Harvesting microalgal biomass using submerged microfiltration membranes, Bioresour. Technol. 111 (2012) 343–352.
[55] R. Honda, J. Boonnorat, C. Chiemchaisri, W. Chiemchaisri, K. Yamamoto, Carbon dioxide capture and nutrients removal utilizing treated sewage by concentrated microalgae cultivation in a membrane photobioreactor, Bioresour. Technol. 125 (2012) 59–64.
[56] H.T. Tse, S. Luo, J. Li, Z. He, Coupling microbial fuel cells with a membrane photobioreactor for wastewater treatment and bioenergy production, Bioprocess Biosyst. Eng. 39 (11) (2016) 1703–1710.
[57] X. Li, W. Li, J. Zhai, H. Wei, Q. Wang, Effect of ammonium nitrogen on microalgal growth, biochemical composition and photosynthetic performance in mixotrophic cultivation, Bioresour. Technol. 273 (2019) 368–376.
[58] G. Salbitani, S. Carfagna, Ammonium utilization in microalgae: a sustainable method for wastewater treatment, Sustainability 13 (2) (2021).
[59] R. Kakarla, J.R. Kim, B.-H. Jeon, B. Min, Enhanced performance of an air–cathode microbial fuel cell with oxygen supply from an externally connected algal bioreactor, Bioresour. Technol. 195 (2015) 210–216.
[60] N. Rashid, Y.-F. Cui, M.S.U. Rehman, J.-I. Han, Enhanced electricity generation by using algae biomass and activated sludge in microbial fuel cell, Sci. Total Environ. 456–457 (2013) 91–94.
[61] S.B. Velasquez-Orta, T.P. Curtis, B.E. Logan, Energy from algae using microbial fuel cells, Biotechnol. Bioeng. 103 (6) (2009).
[62] F. Ndayisenga, Z. Yu, Y. Yu, C.-H. Lay, D. Zhou, Bioelectricity generation using microalgal biomass as electron donor in a bio-anode microbial fuel cell, Bioresour. Technol. 270 (2018) 286–293.
[63] C. Xu, K. Poon, M.M. Choi, R. Wang, Using live algae at the anode of a microbial fuel cell to generate electricity, Environ. Sci. Pollut. Res. 22 (20) (2015) 15621–15635.
[64] N. Longtin, D. Oliveira, A. Mahadevan, V. Gejji, C. Gomes, S. Fernando, Analysis of Spirulina platensis microalgal fuel cell, J. Power Sources 486 (2021) 229290.
[65] B. Xu, Z. Ge, Z. He, Sediment microbial fuel cells for wastewater treatment: challenges and opportunities, Environ. Sci. Water Res. Technol. 1 (3) (2015) 279–284.
[66] D.R. Bond, D.E. Holmes, L.M. Tender, D.R. Lovley, Electrode-reducing microorganisms that harvest energy from marine sediments, Science 295 (5554) (2002) 483–485.
[67] C.E. Reimers, L.M. Tender, S. Fertig, W. Wang, Harvesting energy from the marine sediment-water interface, Environ. Sci. Technol. 35 (1) (2001) 192–195.
[68] A.S. Commault, G. Lear, R.J.W. Novis, Photosynthetic biocathode enhances the power output of a sediment-type microbial fuel cell, N. Z. J. Bot. 52 (1) (2014) 48–59.
[69] X. Song, W. Wang, X. Cao, Y. Wang, L. Zou, X. Ge, Y. Zhao, Z. Si, Y. Wang, Chlorella vulgaris on the cathode promoted the performance of sediment microbial fuel cells for electrogenesis and pollutant removal, Sci. Total Environ. 728 (2020) 138011.
[70] Y. Kim, B.E. Logan, Microbial desalination cells for energy production and desalination, Desalination 308 (2013) 122–130.
[71] B. Kokabian, R. Smith, J.P. Brooks, V.G. Gude, Bioelectricity production in photosynthetic microbial desalination cells under different flow configurations, J. Ind. Eng. Chem. 58 (2018) 131–139.
[72] S. Shamil, G. Farooq, T. Abbas, Photosynthetic microbial desalination cell to treat oily wastewater using microalgae Chlorella vulgaris, Civ. Eng. J. 5 (2019) 2686–2699.
[73] B. Kokabian, V.G. Gude, Photosynthetic microbial desalination cells (PMDCs) for clean energy, water and biomass production, Environ. Sci. Process Impacts 15 (12) (2013) 2178–2185.
[74] A. Khandelwal, M. Chhabra, P. Yadav, Performance evaluation of algae assisted microbial fuel cell under outdoor conditions, Bioresour. Technol. 310 (2020) 123418.

[75] C.N. Reddy, M.P. Sudhakar, B. Min, P. Shanmugam, Future perspectives on cost-effective microbial fuel cells in rural areas, in: V. Sivasankar, P. Mylsamy, K. Omine (Eds.), Microbial Fuel Cell Technology for Bioelectricity, Springer Publishing, New York, 2018, pp. 283–302.
[76] W.-W. Li, H.-Q. Yu, Z. He, Towards sustainable wastewater treatment by using microbial fuel cells-centered technologies, Energy Environ. Sci. 7 (3) (2014) 911–924.
[77] P. Liang, R. Duan, Y. Jiang, X. Zhang, Y. Qiu, X. Huang, One-year operation of 1000-L modularized microbial fuel cell for municipal wastewater treatment, Water Res. 141 (2018) 1–8.
[78] Z. Yang, L. Zhang, C. Nie, Q. Hou, S. Zhang, H. Pei, Multiple anodic chambers sharing an algal raceway pond to establish a photosynthetic microbial fuel cell stack: voltage boosting accompany wastewater treatment, Water Res. 164 (2019) 114955.

CHAPTER 2

Algal intervention as nature-based solution for treatment of landfill leachate

Sandeep K. Malyan[a], Sumit Kumar[b], Rajesh Singh[c], Sandeep Singh[c], Gagan Anand[d], Shefali Upadhyay[e], Kajal Saini[e], and Smita S. Kumar[e,f]
[a]Department of Environmental Studies, Dyal Singh Evening College, University of Delhi, New Delhi, India
[b]Centre for Rural Development & Technology, Indian Institute of Technology Delhi, Hauz Khas, Delhi, India
[c]Environmental Hydrology Division, National Institute of Hydrology, Roorkee, India
[d]Applied Science Cluster, SOAE, University of Petroleum and Energy Studies, Dehradun, India
[e]Department of Environmental Sciences, J.C. Bose University of Science & Technology, YMCA, Faridabad, Haryana, India
[f]Department of Environmental Studies, Central University of Haryana, Jant-Pali, Mahendergarh, Haryana, India

1 Introduction

The generation of different types of solid waste and its management is among the prime concerns of the 21st century. Waste generation is rapidly increasing with urbanization and modernization. Sanitary landfills are scientifically engineered designs meant for managing problems associated with the dumping of MSW (Fig. 1). Unscientific dumping of this waste can result in air, water, and soil pollution. Unplanned urbanization is resulting in a high volume of waste generation and is a global menace right now. Ahluwali and Patel [1] quoted that the volume of waste generation may increase by up to 125 million tons by 2031 in India. Dutta and Jinsart [2] quoted that in India only 12.5% of the total generated waste is managed scientifically and the rest is dumped in open land around the cities. The dumping of 87.5% of waste in open dumpsites is alarming and hinders the goal of environmental sustainability. The untreated municipal solid waste (MSW) is a mixture of dry and wet as well as toxic and nontoxic waste.

Leachate is produced from waste dumped in landfills and its quality and composition (Table 1) varies with the age of landfill and the nature of the waste (Fig. 2). Landfill leachate (LL) contains inorganic substances [3,4], organic compounds [5,6], microplastics [7–9], pesticides [10–13], and other emerging toxic pollutants [14–16]. The decomposition of organic waste in the anaerobic environment of the landfills generates methane (CH_4) a potent greenhouse gas in the atmosphere [17–19]. Methane is 25 times higher potent greenhouse gas [20] as compared to carbon dioxide and landfills release a significant quantity of CH_4 to the atmosphere thus contributing to global warming. Removal of these pollutants from LL is essential before discharging for environmental protection. Chemical, physical, and biological technologies can be used for the remediation of these

Algae Based Bioelectrochemical Systems for Carbon Sequestration, Carbon Storage, Bioremediation and Bioproduct Generation
https://doi.org/10.1016/B978-0-323-91023-1.00014-4

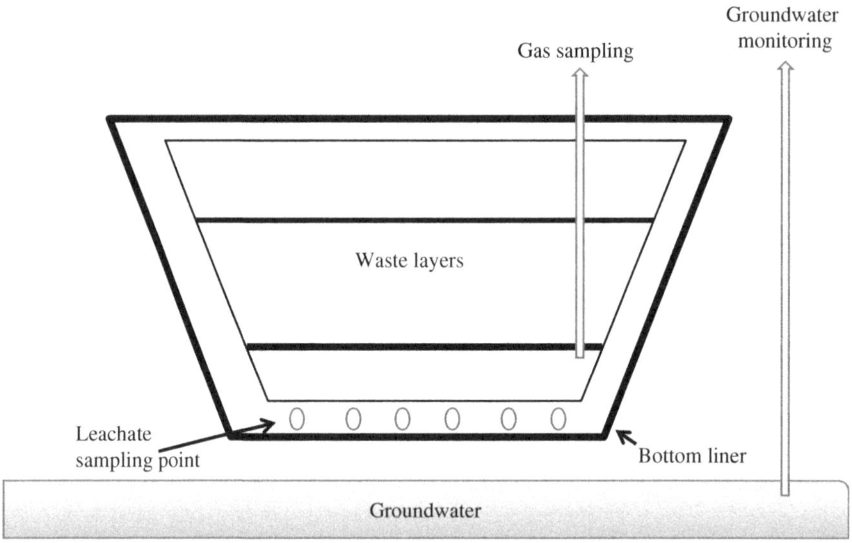

Fig. 1 Design of landfill with different components.

pollutants from LL (Table 2). If the BOD/COD ratio is high, biological treatment technologies are desirable. On the other hand, if the concentration of xenobiotic compounds is higher as in the case of mature leachates, physicochemical approaches are employed. Physicochemical methods reported for the treatment of landfill leachate include adsorption, advanced oxidation, chemical precipitation, coagulation–flocculation, electrochemical treatment, ion exchange, membrane filtration, etc. Generally, the concentration of biochemical oxygen demand and chemical oxygen demand is high in LL and a high dose of chemicals is required to tackle this issue which also enhances the economic cost of the process in addition to the additional chemical load. Biological technologies such as constructed wetlands, photobioreactors, algal treatment, etc., can also be used for the mitigation of pollution load of the LL [20,36,37]. Pollutants such as nitrogen, phosphorus, and trace metals are nutrients for plants, algae, and several microbes up to a certain level. In this chapter, the removal of pollutants from LL has been discussed through algae-based interventions.

2 Biological approaches for the landfill leachate treatment

Landfill leachate is highly contaminated heterogonous liquid noxious waste and its treatment is necessary before the final discharge. Physical, chemical, and biological treatment methods are widely adopted for the treatment of landfill leachate. However, biological approaches are considered cost-effective and environmentally sound as well as sustainable. Constructed wetlands, microbial fuel cells (MFC), micro-algal consortium, etc., are some of

Table 1 Concentration of various leachate constituents in the different stages of biological stabilization.

Leachate constituent	Phase of biological stabilization			
	Transition phase	Acid formation methane	Fermentation	Final maturation
(Chemical oxygen demand) COD	480–18,000	1500–71,100	580.0–9760	31.00–900.0
Nitrate	0.100–51.00	0.050–19.00	–	0.500–0.600
Ammoniacal nitrogen (NH_3-N)	120–225.0	2.000–1030	6.000–430.0	–
Phosphate	0.600–1.700	0.200–120.0	0.700–14.00	0.200–14.00
Total solids (TS)	2450–2960	4120–55,300	2090–6410	1460–4640
pH	6.000–7.000	4.700–7.700	6.300–8.800	7.100–8.800
Copper (Cu)	0.085–0.390	0.005–2.200	0.030–0.180	0.020–0.560
Iron (Fe)	68.00–312.0	90–2200	115.0–336.0	4.000–20.00
Lead (Pb)	0.001–0.004	0.010–1.000	0.010–0.100	0.010–0.100
Magnesium (Mg)	60.00–96.00	3–1140	81.00–505.0	81.00–190.0
Manganese (Mn)	0.600	0.600–41.00	0.600	0.600
Potassium (K)	35.00–2300	35.00–2300	35.00–2300	35.00–2300
Sodium (Na)	20.00–7600	–	–	–
Zinc (Zn)	0.060–21	0.650–220.0	0.400–6.00	0.400
Coliform	$10^0.0–10^5.0$	$10^0.0–10^5.0$	–	–
Electrical conductivity (EC)	2450–3310	1600–17,100	2900–7700	1400–4500
Chloride (Cl)	30.00–5000	30.00–5000	30.00–5000	30.00–5000
Sulfate	10.00–458.0	10.00–3240	–	5.000–40.00
Sulfide	–	0.000–818.0	0.900	–
Cadmium (Cd)	<0.005–0.01	<0.005–0.39	<0.005–0.10	0.004
Calcium (Ca)	190.0–490.0	70.00–3900	76.00–490.0	76–254.0
Chromium (Cr)	0.023–0.280	0.060–18.00	0.050	0.050

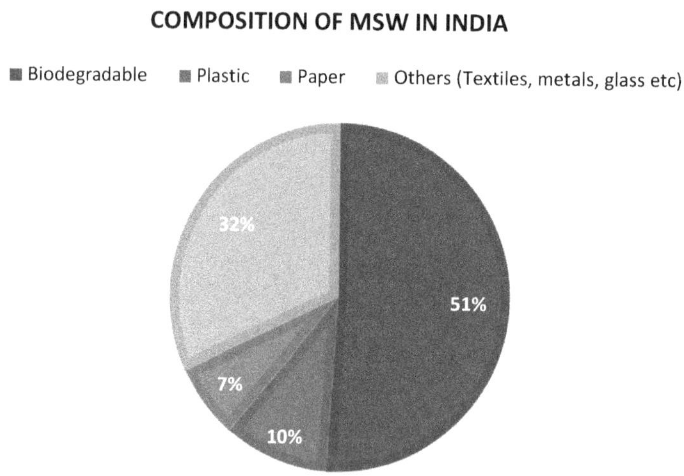

Fig. 2 Composition of municipal solid waste in India. *(Source: I.J. Ahluwalia, U. Patel, Solid Waste Management in India: An Assessment of Resource Recovery and Environmental Impact, 2018.)*

the most popular and widely adopted biological technologies for the treatment of landfill leachate (Table 3). The only limitation is that several toxic contaminants present in the leachate can negatively affect the microbial biocatalysts and suppress their removal efficiency. Biological methods in combination with chemical or physical treatment technologies such as pretreatment, posttreatment, and simultaneous treatment methods can have more significant output and can reduce the negative impacts [20,52].

3 Algae and landfill leachate remediation

Several algal species such as *Galdieria sulphuraria*, *Chlorella pyrenoidosa*, *Ankistrodesmus*, *Nostoc*, *Anabaena*, *Chlorella vulgaris*, *Synedra acus*, *Spirulina* sp., *Trichocerca* sp., *Closteriopsis longissima*, with *Euglena* sp., etc., have great bioremediation potential for landfill leachate (Table 4). The bioremediation potential of algal species *C. pyrenoidosa* has been reported by Li et al. [66] in China. Pan et al. [60] investigated the nitrogen removal efficiency of *Galdieria sulphuraria* for real landfill leachate using 20%, 40%, 80%, and 100% dilution factors (Table 4). The best algal growth rate (159 mg/L day) and biomass density (2.16 g/L) were observed at 20% of the dilution which clearly indicated that nutrient removal through algal treatment is an economical option [60]. The algal biomass density gets completely inhibited at 40% or higher landfill leachate concentration due to the high concentration of the heavy metals [60], and generally, it is recommended to dilute landfill leachate at least five times [60]. A similar finding is also reported by Nair and Nagendra [64] for the removal of total nitrogen, dissolved organic carbon, and other pollutants through *C. pyrenoidosa* (Table 4).

Table 2 Landfill leachate treatment options [21].

Treatment technology	Conditions used	BOD (mg/L)	COD (mg/L)	NH_4^+-N (mg/L)	Removal efficiency	References
Coagulation/ membrane filtration	Integration of ultrafiltration membrane process with chemical coagulation	ND	ND	ND	Turbidity 73%	[22]
AOP/adsorption (ion exchange)	Supercritical water oxidation (ScWO)/zeolite	134	1255	355	COD 74%	[23]
AOP	Vermiculite/ozonation	ND	ND	ND	COD 17%	[24]
AOP/coagulation	Electrooxidation and coagulation	2098	3181	963	COD 90% + Ammonia 90% +	[25]
Coagulation/ flocculation	Tannin-based natural coagulant	59	893	531	COD 55% Ammonia 91%	[26]
AOP	Fenton process	ND	10,516	ND	COD 79%	[27]
Coagulation/ flocculation	Polyaluminum chloride and *Dimocarpus longan* seeds as flocculant	130	3036	794	COD 62%	[28]
Coagulation/ flocculation	Red earth as a coagulant	ND	ND	ND	COD 67% Ammonia 43%	[29]
AOP/coagulation flocculation	UV-based sulfate radical oxidation process/ coagulation–flocculation	351	5123	2700	COD 91% Ammonia 5%	[30]
AOP	Ozone/catalyst (ZrCl4)	234	3125	1674	COD 88% Ammonia 79%	[31]
Adsorption	Silica nanoparticle	ND	19,691	ND	COD 77%	[32]
Constructed wetland	Using *Colocasia esculenta*, *Gynerium sagittatum*, and *Heliconia psittacorum*	ND	626	ND	COD 67%	[33]
Aerobic reactor	Aerobic sequencing batch reactor (ASBR)	ND	3200	1800	COD 43% Ammonia 96%	[34]
AOP/adsorption	MAC/ozonation	ND	ND	ND	COD 74%	[35]

Table 3 Biological treatment methods reported for the treatment of landfill leachate.

Name of the technology	Name of the pollutant removed from landfill leachate	References
Constructed wetlands	Total nitrogen, total phosphorus, COD	[36,38–40]
	Microbial pollutant and nitrogen	[39,41,42]
	Heavy metals	[43,44]
	Pesticides	[13,45]
	Pharmaceuticals and personal care products	[45–47]
	Antibiotic genes	[48]
Microbial fuel cells	BOD, COD, and nutrients	[10,11,20,38,49-51,48]
	Antibiotic genes and personal care products	[48]
Microalgal consortium	BOD, COD, and nutrients	[52,53]
	Pesticides and heavy metals	[54]
Algal photobioreactors	Nutrients, BOD and COD	[55]

Integration of microalgae with other treatment technologies can considerably improve the process. The integration of ozonation with microalgal treatment removed 99.7% of ammoniacal nitrogen [41,42]. The pretreatment of landfill leachate through the ozonation process reduced the chromaticity significantly which further enhances the nitrogen removal efficiency of the microalgae [41,42] (Table 4). Khanzada and Övez [65] investigated the potential of *Chlamydomonas reinhardii* and *C. vulgaris* microalgae for the bioremediation of filtrated landfill leachate (Table 4). *Chlamydomonas reinhardii* and *C. vulgaris* removed ammonium with 100% efficiency from a 10% diluted landfill leachate having 50 mg/L of ammonium, while higher dilution showed lower ammonium nitrogen removal [65]. Khanzada [62] investigated the removal potential of microalgae (*Chlamydomonas reinhardii* and *C. vulgaris*) for the ammonium nitrogen and phosphorus (Table 4). Microalgae removes phosphorus more quickly as compared to ammonium/nitrate nitrogen from landfill leachate [62]. Microalgae take up phosphorus in intracellular poly-P granules form which results in the significant removal of phosphorus from wastewater [62]. Removal of dissolved organic carbon (81%), total nitrogen (70%), and orthophosphate through microalgae are also reported by Nair and Nagendra [64] (Table 4).

On the basis of the above findings, it can be concluded that landfill leachate is a toxic pollutant that can significantly be treated biologically with microalgae, provided the dilution is optimum dilution and/or the leachate has been pretreated to reduce the concentration of pollutants.

Table 4 Removal of different contaminants from landfill leachate through algae-based technology.

Reference	Location of landfill leachate used in the study	Name of algal species	Gene Bank accession No	Pollutant removal	Remarks
[56]	Blang Bintang Aceh Besar, Indonesia	*Paramecium* sp., *Synedra acus*, *Spirulina* sp., *Trichocerca* sp., *Closteriopsis longissima*, with *Euglena* sp., and *Euglena* sp.	–	Organic and nitrogen compounds	Six different microalgae species were used
[41,42]	Maiyuan landfill in Nanchang, China	*Chlorella* sp.	–	Nitrogen	Integrated approach of ozonation and microalgae
[55]	Perungudi, dumpsite of Chennai, India.	*Chlorella pyrenoidosa* and *Chlorella vulgaris*	–	TN, TP, and heavy metals	Integrated approaches (microwave and algal photobioreactors)
[57]	Saravan landfill, Guilan Province, Iran	*Chlorella* sp.	–	Phosphate, ammonium, and nitrate	Nutrients removal from LL and their use in poor nutrient-poor agricultural soil
[58]	Sanwei Environmental Protection Co., Ltd. (Chongqing, China)	*C. vulgaris*	–	Nutrients	Nutrient removal from LL followed by dark fermentation for bioenergy production
[59]	Commercial dump site of Islamabad, Pakistan	*Ankistrodesmus*, *Nostoc*, and *Anabaena*	–	Heavy metals	Green algae and cyanobacteria were explored for the removal of heavy metals from LL

Continued

Table 4 Removal of different contaminants from landfill leachate through algae-based technology—cont'd

Reference	Location of landfill leachate used in the study	Name of algal species	Gene Bank accession No	Pollutant removal	Remarks
[60]	St. John the Baptis, LL, Beaumont, USA	*Galdieria sulphuraria*	CCMEE 5587.1	Nutrients	Investigated the on-site landfill leachate treatment
[61]	Landfill in northern Portugal	*C. vulgaris* and *Tetradesmus obliquus*	CCAP 211/11B and CCAP 276/3A	Nitrate and nitrite	Pretreated LL was treated with microalgae
[62]	Istanbul municipal landfill management, Istanbul, Turkey	*Chlamydomonas reinhardii* and *C. vulgaris*	–	Phosphorus and ammonium nitrogen	Nitrogen and phosphorus removal was investigated
[63]	Dumpsites of Chennai, India	*C. pyrenoidosa*	NCIM 2738	DOC, TN, NH_4-N, and phosphates	Pretreated LL (with electrocoagulation) was biologically treated with microalgae
[64]	Dumpsite in Chennai, India	*C. pyrenoidosa*	–	TN, DOC, and orthophosphate	Phycoremediation potential of algal species is reported
[54]	Four different landfill sites in Delhi, India	*Scenedesmus* sp. ISTGA1	KC 342184	Heavy metals, pesticides, and organic compounds	Integrated approaches using bacteria and algae
[65]	Istanbul municipal landfill management, Istanbul, Turkey	*Chlamydomonas reinhardii* and *C. vulgaris*	–	Ammonium nitrogen	Microalgae used for ammonium nitrogen removal from LL

4 Conclusion

Landfill leachate is a heterogeneous toxic waste produced during the decomposition of solid waste in landfills or dumpsites. Landfill leachate is a mixture of many pollutants such as dissolved organics, inorganic compounds, heavy metals, and xenobiotics. Due to the prevalence of all these contaminants, landfill leachate must be treated to save the environment. The chemical treatment approaches generate byproducts that are toxic to the environment. Therefore, biological treatment is preferred due to economic feasibility and environmental sustainability. Algae need several compounds and several trace metals present in landfill leachate as nutrients (such as nitrate, ammonium, and phosphorus) for growth and development. However, the high concentration of other toxic contaminants can suppress the algal growth if raw landfill leachate is fed to algae as such. To overcome the negative effects, landfill leachate needs to be diluted or pretreated to facilitate the growth of microalgae. Several other by-products such as biofuel can also be further harvested from the algal biomass. The efficiency of pollutant removal such as nitrogen, phosphorus, etc., from the landfill leachate is affected by dilution and type of algae species. The integration of algal remediation with constructed wetlands or some suitable chemical treatment technology also has significance in terms of environmental sustainability.

Acknowledgments

The author (S.K.M.) gratefully acknowledges the Dyal Singh Evening College, University of Delhi, New Delhi, India, for all the necessary support.

References

[1] I.J. Ahluwalia, U. Patel, Solid Waste Management in India: An Assessment of Resource Recovery and Environmental Impact, 2018.
[2] A. Dutta, W. Jinsart, Waste generation and management status in the fast-expanding Indian cities: a review, J. Air Waste Manage. Assoc. 70 (5) (2020) 491–503.
[3] M.J. Quina, C.T. Pinheiro, Inorganic waste generated in Kraft pulp mills: the transition from landfill to industrial applications, Appl. Sci. 10 (7) (2020) 2317.
[4] X. Zou, A. Mohammed, M. Gao, Y. Liu, Mature landfill leachate treatment using granular sludge-based reactor (GSR) via nitritation/denitritation: process startup and optimization, Sci. Total Environ. 157078 (2022).
[5] S.A. Siddiqi, A. Al-Mamun, A. Sana, M.S. Baawain, M.R. Choudhury, Characterization and pollution potential of leachate from urban landfills during dry and wet periods in arid regions, Water Supply 22 (3) (2022) 3462–3483.
[6] P. Wijekoon, P.A. Koliyabandara, A.T. Cooray, S.S. Lam, B.C. Athapattu, M. Vithanage, Progress and prospects in mitigation of landfill leachate pollution: risk, pollution potential, treatment and challenges, J. Hazard. Mater. 421 (2022) 126627.
[7] P. He, L. Chen, L. Shao, H. Zhang, F. Lü, Municipal solid waste (MSW) landfill: a source of microplastics?-evidence of microplastics in landfill leachate, Water Res. 159 (2019) 38–45.
[8] L. Hou, D. Kumar, C.G. Yoo, I. Gitsov, E.L.W. Majumder, Conversion and removal strategies for microplastics in wastewater treatment plants and landfills, Chem. Eng. J. 406 (2021) 126715.

[9] J. Sun, Z.R. Zhu, W.H. Li, X. Yan, L.K. Wang, L. Zhang, B.J. Ni, Revisiting microplastics in landfill leachate: unnoticed tiny microplastics and their fate in treatment works, Water Res. 190 (2021) 116784.

[10] S.S. Kumar, P. Ghosh, S.K. Malyan, J. Sharma, V. Kumar, A comprehensive review on enzymatic degradation of the organophosphate pesticide malathion in the environment, J. Environ. Sci. Health C 37 (4) (2019) 288–329.

[11] S.S. Kumar, V. Kumar, R. Kumar, S.K. Malyan, N.R. Bishnoi, Ferrous sulfate as an in-situ anodic coagulant for enhanced bioelectricity generation and COD removal from landfill leachate, Energy 176 (2019) 570–581.

[12] L. Pisharody, A. Gopinath, M. Malhotra, P.V. Nidheesh, M.S. Kumar, Occurrence of organic micropollutants in municipal landfill leachate and its effective treatment by advanced oxidation processes, Chemosphere 287 (2022) 132216.

[13] C. Yang, B. Wang, H. Wang, Z. He, Y. Pi, J. Zhou, T. Fu, Removal of organochlorine pesticides and metagenomic analysis by multi-stage constructed wetland treating landfill leachate, Chemosphere 301 (2022) 134761.

[14] J. Barasarathi, H. Auwal, A. Pariatamby, F.S. Hamid, E.C. Uche, Phytoremediation of leachate contaminated soil: a biotechnical option for the bioreduction of heavy metals induced pollution in tropical landfill, Environ. Sci. Pollut. Res. 29 (15) (2022) 22069–22081.

[15] S.K. Malyan, R. Singh, M. Rawat, M. Kumar, A. Pugazhendhi, A. Kumar, S.S. Kumar, An overview of carcinogenic pollutants in groundwater of India, Biocatal. Agric. Biotechnol. 21 (2019) 101288.

[16] J. Torkashvand, K. Godini, S. Norouzi, M. Gholami, M. Yeganeh, M. Farzadkia, Effect of cigarette butt on concentration of heavy metals in landfill leachate: health and ecological risk assessment, J. Environ. Health Sci. Eng. 19 (1) (2021) 483–490.

[17] S.K. Malyan, A. Bhatia, A. Kumar, D.K. Gupta, R. Singh, S.S. Kumar, N. Jain, Methane production, oxidation and mitigation: a mechanistic understanding and comprehensive evaluation of influencing factors, Sci. Total Environ. 572 (2016) 874–896.

[18] S.K. Malyan, O. Singh, A. Kumar, G. Anand, R. Singh, S. Singh, A. Kumar, Greenhouse gases tradeoff from ponds: an overview of emission process and their driving factors, Water 14 (6) (2022) 970.

[19] I.D.A.A. Warmadewanthi, G. Chrystiadini, S.B. Kurniawan, S.R.S. Abdullah, Impact of degraded solid waste utilization as a daily cover for landfill on the formation of methane and leachate, Bioresour. Technol. Rep. 15 (2021) 100797.

[20] S.K. Malyan, S.S. Kumar, L. Singh, R. Singh, D.A. Jadhav, V. Kumar, Bioelectrochemical systems for removal and recovery of heavy metals, in: Bioremediation, Nutrients, and Other Valuable Product Recovery, Elsevier, 2021, pp. 185–203.

[21] L.M.L.K.B. Lindamulla, N.K.R.N. Jayawardene, W.S.M.S.K. Wijerathne, M. Othman, K.G.N. Nanayakkara, K.B.S.N. Jinadasa, G.B.B. Herath, V. Jegatheesan, Treatment of mature landfill leachate in tropical climate using membrane bioreactors with different configurations, Chemosphere 307 (2022) 136013, https://doi.org/10.1016/j.chemosphere.2022.136013.

[22] S. Nazia, N. Sahu, V. Jegatheesan, S.K. Bhargava, S. Sridhar, Integration of ultrafiltration membrane process with chemical coagulation for proficient treatment of old industrial landfill leachate, Chem. Eng. J. 412 (2021) 128598.

[23] A.P.J. Scandelai, J.P. Zotesso, V. Jegatheesan, L. Cardozo-Filho, C.R.G. Tavares, Intensification of supercritical water oxidation (ScWO) process for landfill leachate treatment through ion exchange with zeolite, Waste Manag. 101 (2020) 259–267.

[24] W.L.M. Braga, et al., Optimization of the treatment of sanitary landfill by the ozonization catalysed by modified nanovermiculite in a rotating packed bed, J. Clean. Prod. 249 (2020) 119395.

[25] M.S. de Oliveira, L.F. da Silva, A.D. Barbosa, L.L. Romualdo, G. Sadoyama, L.S. Andrade, Landfill leachate treatment by combining coagulation and advanced electrochemical oxidation techniques, Chem. Electro. Chem. 6 (5) (2019) 1427–1433.

[26] T.J. Banch, M.M. Hanafah, A.F. Alkarkhi, S.S. Abu Amr, Factorial design and optimization of landfill leachate treatment using tannin-based natural coagulant, Polymers 11(8)1349 (2019).

[27] A. Maslahati Roudi, S. Chelliapan, W.H.M. Wan Mohtar, H. Kamyab, Prediction and optimization of the Fenton process for the treatment of landfill leachate using an artificial neural network, Water 10 (5) (2018) 595.

[28] H.A. Aziz, N.A. Rahim, S.F. Ramli, M.Y. Alazaiza, F.M. Omar, Y.-T. Hung, Potential use of Dimocarpus longan seeds as a flocculant in landfill leachate treatment, Water 10 (11) (2018) 1672.

[29] N.A. Zainol, L.B. Pin, N.A. Rashid, A.A. Ghani, S.N. Zailani, A.L.A. Rani, Treatment of landfill leachate by coagulation flocculation process using red earth as coagulant, AIP Conf. Proc. 2030 (1) (2018) 020043. AIP Publishing LLC.

[30] A.R. Ishak, F.S. Hamid, S. Mohamad, K.S. Tay, Stabilized landfill leachate treatment by coagulation-flocculation coupled with UV-based sulfate radical oxidation process, Waste Manag. 76 (2018) 575–581.

[31] S.S.A. Amr, S.N.F. Zakaria, H.A. Aziz, Performance of combined ozone and zirconium tetrachloride in stabilized landfill leachate treatment, J. Mater. Cycles Waste Manag. 19 (4) (2017) 1384–1390.

[32] S. Pavithra, S. Shanthakumar, Removal of COD, BOD and color from municipal solid waste leachate using silica and iron nanoparticles–a comparative study, Global Nest J. 19 (2017) 122–130.

[33] C.A. Madera-Parra, Treatment of landfill leachate by polyculture constructed wetlands planted with native plants, Ing. Compet. 18 (2) (2016) 183–192.

[34] C.K. Lim, et al., Treatment of landfill leachate using ASBR combined with zeolite adsorption technology, 3 Biotech 6 (2) (2016) 1–6.

[35] J.-M. Wang, C.-S. Lu, Y.-Y. Chen, H.-J. Fan, Landfill leachate treatment with Mn and Ce oxides impregnated GAC–ozone treatment process, Colloids Surf. A Physicochem. Eng. Asp. 482 (2015) 536–543.

[36] T. Saeed, M.J. Miah, N. Majed, M.K. Alam, T. Khan, Effect of effluent recirculation on nutrients and organics removal performance of hybrid constructed wetlands: landfill leachate treatment, J. Clean. Prod. 282 (2021) 125427.

[37] S.K. Malyan, S. Yadav, V. Sonkar, V.C. Goyal, O. Singh, R. Singh, Mechanistic understanding of the pollutant removal and transformation processes in the constructed wetland system, Water Environ. Res. 93 (10) (2021) 1882–1909, https://doi.org/10.1002/wer.1599.

[38] T. Saeed, N. Majed, M.J. Miah, A.K. Yadav, A comparative landfill leachate treatment performance in normal and electrodes integrated hybrid constructed wetlands under unstable pollutant loadings, Sci. Total Environ. 155942 (2022).

[39] I. Arliyani, B.V. Tangahu, S. Mangkoedihardjo, Performance of reactive nitrogen in leachate treatment in constructed wetlands, J. Ecol. Eng. 22 (5) (2021).

[40] A. Ikhlaq, F. Javed, A. Akram, U.Y. Qazi, Z. Masood, T. Ahmed, F. Qi, Treatment of leachate through constructed wetlands using Typha angustifolia in combination with catalytic ozonation on Fe-zeolite a, Int. J. Phytoremediation 23 (8) (2021) 809–817.

[41] H. Wang, C. Yang, B. Wang, Z. He, T. Fu, Nitrogen removal performance and microbiological characteristics for the landfill leachate treatment in a three-stage vertical flow constructed wetlands system, Environ. Technol. Innov. (2022) 102728.

[42] Z. Wang, Z. Xiong, L. Yang, L. Lai, H. Xiao, Y. Ding, X. Luo, Enhancing nitrogen removal in mature landfill leachate by mixed microalgae through elimination of inhibiting factors, Sci. Total Environ. 828 (2022) 154530.

[43] M.A. Maine, H.R. Hadad, N.E. Camaño Silvestrini, E. Nocetti, G.C. Sanchez, M.A. Campagnoli, Cr, Ni, and Zn removal from landfill leachate using vertical flow wetlands planted with Typha domingensis and Canna indica, Int. J. Phytoremediation 24 (1) (2022) 66–75.

[44] A. Wdowczyk, A. Szymańska-Pulikowska, B. Gałka, Removal of selected pollutants from landfill leachate in constructed wetlands with different filling, Bioresour. Technol. 353 (2022) 127136.

[45] C. Yang, T. Fu, H. Wang, R. Chen, B. Wang, T. He, M. Chen, Removal of organic pollutants by effluent recirculation constructed wetlands system treating landfill leachate, Environ. Technol. Innov. 24 (2021) 101843.

[46] Y. Wang, T. Yin, B.C. Kelly, K.Y.H. Gin, Bioaccumulation behaviour of pharmaceuticals and personal care products in a constructed wetland, Chemosphere 222 (2019) 275–285.

[47] X. Yi, N.H. Tran, T. Yin, Y. He, K.Y.H. Gin, Removal of selected PPCPs, EDCs, and antibiotic resistance genes in landfill leachate by a full-scale constructed wetlands system, Water Res. 121 (2017) 46–60.

[48] H.-L. Song, H. Li, S. Zhang, Y.-L. Yang, L.-M. Zhang, H. Xu, X.-L. Yang, Fate of sulfadiazine and its corresponding resistance genes in up-flow microbial fuel cell coupled constructed wetlands: effects of

circuit operation mode and hydraulic retention time, Chem. Eng. J. 350 (2018) 920–929, https://doi.org/10.1016/j.cej.2018.06.035.
[49] K. Elmaadawy, B. Liu, G.K. Hassan, X. Wang, Q. Wang, J. Hu, X. Wu, Microalgae-assisted fixed-film activated sludge MFC for landfill leachate treatment and energy recovery, Process Saf. Environ. Prot. 160 (2022) 221–231.
[50] P.K. Pandis, T. Kamperidis, K. Bariamis, I. Vlachos, C. Argirusis, V.N. Stathopoulos, A. Tremouli, Comparative study of different production methods of activated carbon cathodic electrodes in single chamber MFC treating municipal landfill leachate, Appl. Sci. 12 (6) (2022) 2991.
[51] S.S. Kumar, S.K. Malyan, N.R. Bishnoi, Performance of buffered ferric chloride as terminal electron acceptor in dual chamber microbial fuel cell, J. Environ. Chem. Eng. 5 (1) (2017) 1238–1243.
[52] I. Dogaris, E. Ammar, G.P. Philippidis, Prospects of integrating algae technologies into landfill leachate treatment, World J. Microbiol. Biotechnol. 36 (3) (2020) 1–25.
[53] K.D. Sniffen, C.M. Sales, M.S. Olson, The fate of nitrogen through algal treatment of landfill leachate, Algal Res. 30 (2018) 50–58.
[54] M. Kumari, P. Ghosh, I.S. Thakur, Landfill leachate treatment using bacto-algal co-culture: an integrated approach using chemical analyses and toxicological assessment, Ecotoxicol. Environ. Saf. 128 (2016) 44–51, https://doi.org/10.1016/j.ecoenv.2016.02.009.
[55] B.K. Tripathy, M. Kumar, Leachate treatment using sequential microwave and algal photo-bioreactor: effect of pretreatment on reactor performance and biomass productivity, J. Environ. Manag. 311 (2022) 114830.
[56] N. Emalya, L. Mairiza, P.Z. Bilqis, S. Suhendrayatna, E. Munawar, Y. Yunardi, Removal of organic and nitrogen compounds from domestic landfill leachate by microalgae, Biointerface Res. Appl. Chem. 13 (2023), https://doi.org/10.33263/BRIAC132.131.
[57] E. Saadat, N. Ghorbanzadeh, M.B. Farhangi, M. Fazeli Sangani, Potential application of Chlorella sp. biomass cultivated in landfill leachate as agricultural fertilizer, Arch. Agron. Soil Sci. (2022) 1–16.
[58] H. Feng, C. Sun, C. Zhang, H. Chang, N. Zhong, W. Wu, S.H. Ho, Bioconversion of mature landfill leachate into biohydrogen and volatile fatty acids via microalgal photosynthesis together with dark fermentation, Energy Convers. Manag. 252 (2022) 115035.
[59] J. Iqbal, A. Javed, M.A. Baig, Heavy metals removal from dumpsite leachate by algae and cyanobacteria, Bioremediat. J. 26 (1) (2022) 31–40, https://doi.org/10.1080/10889868.2021.1884530.
[60] S. Pan, K.L. Dixon, T. Nawaz, A. Rahman, T. Selvaratnam, Evaluation of Galdieria sulphuraria for nitrogen removal and biomass production from raw landfill leachate, Algal Res. 54 (2021) 102183.
[61] B. Porto, A.L. Gonçalves, A.F. Esteves, S.M.G.U. de Souza, A.A. de Souza, V.J. Vilar, J.C. Pires, Assessing the potential of microalgae for nutrients removal from a landfill leachate using an innovative tubular photobioreactor, Chem. Eng. J. 413 (2021) 127546.
[62] Z.T. Khanzada, Phosphorus removal from landfill leachate by microalgae, Biotechnol. Rep. 25 (2020) e00419.
[63] A.T. Nair, J. Senthilnathan, S.S. Nagendra, Application of the phycoremediation process for tertiary treatment of landfill leachate and carbon dioxide mitigation, J. Water Process. Eng. 28 (2019) 322–330.
[64] A.T. Nair, S.S. Nagendra, *Chlorella pyrenoidosa* mediated phycoremediation of landfill leachate, in: Impact of Global Atmospheric Changes on Natural Resources, 2018, pp. 65–68.
[65] Z.T. Khanzada, S. Övez, Microalgae as a sustainable biological system for improving leachate quality, Energy 140 (2017) 757–765.
[66] S. Li, P.L. Show, H.H. Ngo, S.-H. Ho, Algae-mediated antibiotic wastewater treatment: a critical review, Environ. Sci. Ecotechnol. 9 (2022) 100145.

CHAPTER 3

Progress and prospects of algae-based microbial fuel cells

Jyoti Sharma[a], Manoj Kumar[b], Aarti Ahlawat[c], Shewta Mishra[d], Smita S. Kumar[e,f], and Preeti Sharma[g]
[a]Central Pollution Control Board, New Delhi, India
[b]Department of Geography, Kurukshetra University, Kurukshetra, India
[c]Department of Chemistry, PDM University, Bahadurgarh, Haryana, India
[d]Department of Physics, Birla Institute of Technology and Science, Pilani, India
[e]Department of Environmental Sciences, J.C. Bose University of Science & Technology, YMCA, Faridabad, Haryana, India
[f]Department of Environmental Studies, Central University of Haryana, Jant-Pali, Mahendergarh, Haryana, India
[g]University Institute of Engineering and Technology, Maharshi Dayanand University, Rohtak, India

1 Bioelectrochemical processes

In bioelectrochemical systems, microbes act as catalysts for electrochemical reactions. Microbial electrolysis cell (MEC) and microbial fuel cell (MFC) are the best examples of bioelectrochemical systems. In MECs, external energy is provided at the cathode to produce high-value products, while in MFCs, chemical energy from organic matters is converted into electrical energy. These systems have been extensively employed for the removal of pollutants as well as the recovery of high-value products [1,2]. MFC is an electrochemical device, wherein wastewater is used to generate electric current [2]. Bacteria use organic matter in wastewater to generate energy; however, electron, proton, and CO_2 are produced as by-products. The produced electron passes from anode to cathode through a conductive wire, and the proton passes to the cathode chamber via the proton exchange membrane. Oxygen present at the cathode acts as the oxidant, and combines with hydrogen and electron from the cathode, subsequently completing the circuit and producing water. Researchers found that MFC is an environment-friendly technology for nutrient removal from diverse sources including manure digestate. Yokoyama et al. [3] used MFC to remove nitrogen, potassium, and phosphorus from cow liquid digestate, and the removal up to 84%, 91%, and 70% was achieved. Coupling MFC with constructed in wetland for the treatment of pig manure digestate achieved ammonium removal of 92% [4]. Some researchers coupled MFC with anaerobic digestate (AD) for the removal of nutrients from manure digestate, and found it a good strategy [5]. For the phosphate removal from the anaerobic digestate, MFC–MEC is a promising approach with 90% phosphorus removal [6]. The electron transfer begins at the oxidation reaction at the anode and ends at the reduction reaction at cathode, and microorganisms are used for catalyzing

Algae Based Bioelectrochemical Systems for Carbon Sequestration, Carbon Storage, Bioremediation and Bioproduct Generation
https://doi.org/10.1016/B978-0-323-91023-1.00003-X

the anodic oxidation reaction [2]. MEC as posttreatment of AD has shown good results to reduce and recover ammonium from pig slurry; a two-chamber MEC reactor was used to treat digested pig slurry and achieved a good removal of organics and nitrogen with about 40% ammonium recovery [7]. Sotres et al. [8] investigated the nitrogen recovery from pig slurry by MEC, and the nitrogen flux of $10.3\,gN/dm^2$ was achieved. Besides this, combined ammonia stripping can give a nitrogen recovery of 94.3% due to the high pH in the cathode. A study demonstrated that single-chamber MEC effectively recovered phosphorus by struvite formation at the cathode electrode, which was mainly due to the hydrogen production in the cathode of MEC which increased the localized pH, that was suitable for struvite formation [9]. A novel struvite crystallization method was investigated to recover phosphorus by combining MEC and adding magnesia, and phosphorus recovery of 18%–60% was achieved in the catholyte with different N/P ratios [10].

Microbial catalytic reactions occurring at the anode are responsible for electric power generation from waste and renewable biomass [11,12]. Bioremediation of specific pollutants and nutrients in wastewaters can be done using MFCs [13]. Moreover, MFCs can also be used for the production of bioenergy such as biomethane, biohydrogen, biomass, recovery of heavy metals, decolorization of dyes, etc. [13,14]. Thus, the usage of MFCs provides dual benefits of power generation as well as wastewater treatment, making the process as a whole more environmental friendly and economical [15]. The reduction of oxygen can be catalyzed by abiotic cathode reactions to form water. But, usage of elements like platinum (highly expensive metal) in such processes makes them less economical [11]. Substantial research is being carried out to explore the potential of microalgae for electricity generation in various MFC systems. Bajracharya et al. [16], Butti et al. [17], Singh et al. [18], Freguia et al. [19], and Kelly and He [20] have amiably summarized the different MFC types and discussed the main distinctive characteristics of each system. The use of algae in MFC systems is conducive as algae can be used as efficient electron acceptors during the photosynthetic reactions at the cathodic end or as electron donors at the anodic end of the cell, and thus can effectively remove organic substrates [21,22]. Algae and bacterial community makes a mutual interaction in algal-based MFCs with the benefit of less energy consumption. In the algal-based MFCs, oxidation of biodegradable matter occurs at the anode along with electron generation and CO_2 evolution at the cathode [23]. Various researchers have observed that oxygen production at the cathode mainly depends on the oxygenic photosynthesis for the transfer of electrons from water to NADPþ using the PSI, PSII, and cytochrome *b6f* complex and by small plastoquinone and plastocyanin mobile molecules [21]. In the cathode chamber, algae carry out photosynthesis in the presence of sunlight, and convert CO_2 to generate different types of organic matter, oxygen, and biomass, while in the dark stage, they use up oxygen and produce energy by direct oxidation of the previously produced organic materials [21]. In some cases, as reported by Mohan et al. [14] and Fornero et al. [11], it has been observed that certain photosynthetic cyanobacteria could act as bioanode catalyst for yielding higher electrogenic activity without producing O_2 [24].

The different kinds of MFCs, factors affecting their performance, and wastewater treatment by algae-assisted MFCs have been discussed in subsequent sections.

2 Single-chamber algae MFCs (SCMFCs)

Single-chamber (membraneless) MFCs have been extensively studied till now. Typically, single-chamber MFC consists of two electrodes (the anode and the cathode) connected through an electric circuit (Fig. 1). The photosynthetic biocatalysts can be potentially used in single-chamber MFCs as they have the ability to transport electrons to the electrode surface without having to resort to "electron shuttle mediators" [25,26]. Moreover, some researchers have modified single-chamber algal MFCs wherein both bacteria and algae were added and their collegial interactions increase the efficiency of the process. In this type of cell, microalgae produce organic acids which in turn are used as substrates by bacteria in the presence of light to produce electricity Spirulina platensis, a type of blue–green microalgae, has been studied without using membranes. It was observed that this algae-based MFC system can produce electric power with a power density output of $0.132\,mW\,m^{-2}$ (in the presence of light) and an output of $1.64\,mW\,m^{-2}$ in dark conditions [27]. The results obtained suggested that greater power can be generated in dark conditions rather than

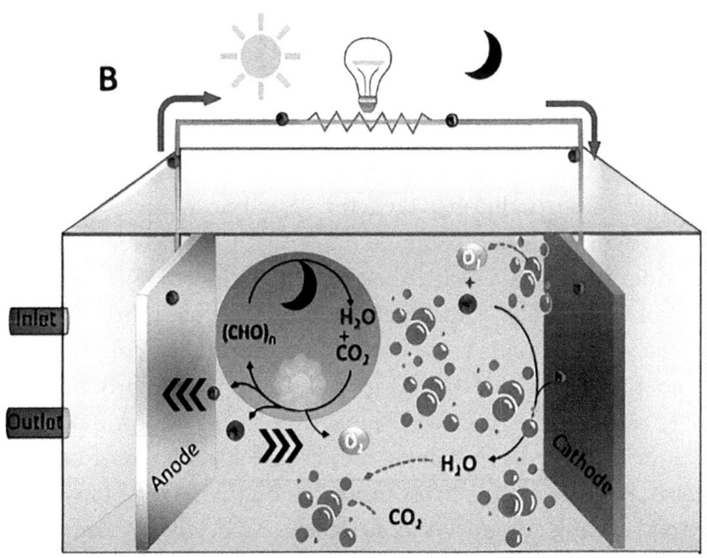

Fig. 1 Single-chamber MFC [25].

under light. With better yields of power under extreme conditions of light, single-chambered MFC has greater durability and prolonged operating time. These MFCs can be practically used as a photosynthetic biocatalyst for bioelectricity generation as their design allows better attachment of microalgae on the surface of the electrodes which in turn increases its efficiency. Nishio et al. [28] have used "single-chamber MFC bioreactors" consisting of algae utilizing Lactobacillus and Geobacter for producing electricity from *Chlamydomonas reinhardtii* grown photosynthetically, and observed that their system generates a power density of $0.078\,W\,m^{-2}$. SCMFCs are the most simplest configurations that allow the power generation using only one chamber.

3 Double-chambered reactor

Double-chambered microbial fuel cells (DCMFCs) have organisms in two different chambers separated by a proton exchange membrane (PEM). This is the most common configuration in algae-assisted MFC studies, owing to the high process efficiencies as shown in Fig. 2. Double-chambered microbial fuel cells can either have the anode, PEM, and cathode encased in one tank or have an H-shaped configuration with the PEM being embedded in the arm. The disadvantages of an shaped-shaped DCMFC include excessive fouling and low membrane area, both of which lead to a decrease in power output. However, for small-scale studies, this is sufficient as the capital cost is reduced.

4 Photosynthetic algal microbial fuel cells (PAMFCs)

Algal growth can be achieved in a variety of bioreactor designs. PAMFCs have a bubble column inside the cathode chamber of an existing DCMFC or couple an algal photobioreactor to a DCMFC. Researchers have demonstrated in a number of studies [29–31] that using a PAMFC as compared to traditional DCMFC can boost power densities and COD removal from a system. Another example of a PAMFC is the technique of combining anaerobic digestion with MFCs [32]. Anaerobic digestion (AD) is the process wherein methanogenesis of volatile fatty acids (VFA) occurs by archaea through the fermentative route in the absence of oxygen. According to Hou et al. [33], coupling AD to bacterial MFCs and algae-assisted cathodes MFCs (AFCs) boosted the production of biogas by 62% and 104%, respectively. Increased rates of an organic compound and nitrogen removal as well as improved pH stabilization are all advantages of combining the AD and AFC systems. However, the power produced by the AD–AFC system was negligible, with a 270 mV starting potential peaking at 490 mV during the procedure. As a result, the AD–AMFC systems are used to increase anaerobic digestion process efficiency but not to produce fuel cell electricity.

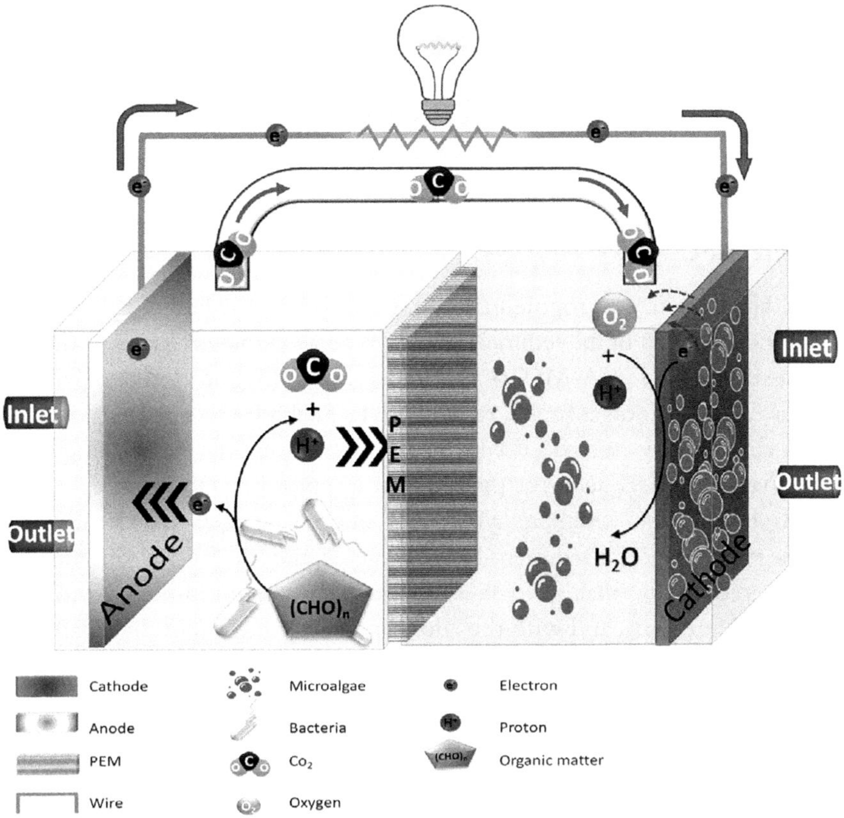

Fig. 2 Dual-chamber algae-assisted MFC [25].

5 Microbial carbon capture cells

In microbial carbon capture (MCC) bioelectrochemical systems, carbon sequestration is achieved by algae present in the cathodic chamber which is fed with off-gas containing carbon dioxide produced in the anodic chamber. The carbon that has been sequestered is then utilized to produce bioelectricity. Microbial carbon capture cells (MCCs) are able to capture carbon dioxide in situ and provide voltage outputs that are better (610 ± 50 mV) than those of MCCs without aeration (630 ± 30 mV). Due to the reduced possibility of carbon dioxide emissions from the anode compartment, MCCs are advantageous over PAMFCs. To reduce the cost of an external CO_2 supply, as well as CO_2 emissions by microalgae and bacteria at the anode and cathode compartments, MCCs can potentially use CO_2 from the ambient or flue gases from the burning and fermentation of fossil fuels [34,35]. By extending the electrode spacing and cathode-specific surface area, MCCs can be scaled up [36]. Due to the increased cathode-specific surface area and electrode spacing (lower solution resistance), they also have higher power densities than MFCs [37].

Additionally, MCCs have been found to achieve high DO concentrations in the cathode compartment up to 7.6mg/L (DO saturation is 7.6mg/L at 30C) [34]. Additionally, the highest power density produced by an air cathode MFC (140mW/m3) is just marginally higher than the figure recorded for a PMMFC (125mW/m^3), indicating that the cost associated with external aeration can be avoided in MFCs made from algae.

6 Sediment MFC

The natural difference in electropotential [38] is created by a sediment buried-anode and immersed cathode on top of the sediment allows to generate bioelectricity in the sediment microalgae-assisted MFCs (MA-MFCs) as shown in Fig. 3. Degradation of organic or inorganic substances by microorganisms is responsible to establish the electron flow, the CO_2 generated is absorbed by microalgal cells in the upper cathodic compartment, and the O_2 that microalgae produce serves as the electron acceptor [39]. Wang et al.'s [40] recent work involved building a sediment MA-MFC using *C. vulgaris*-based cathode which improved the performance of sediment MFC by generating the suitable amount of in situ oxygen that serves as the final electron acceptor. The maximum power output of this MA-MFC was 21mWm^2, and with the addition of a carbon nanotube-coated cathode increased that output by a further 2.4 times. On the other hand, electrogenic microalgae and cyanobacteria can be discovered by using microalgae in the anode side of sediment MA-MFCs [41] Various electrogenic microalgae and cyanobacteria, including Pseudanabaena, Leptolyngbya, Chroococcales, Phormidesmis, Microcoleus, Nostoc, and Phormidium, are linked to the anode. The benthic Chlorophyceae strain Paulschulziapseudovolvox displayed the highest power density in comparison with the other strains [41].

7 Influence of operational parameters on the performance of algae-mediated fuel cells

A lot of factors are responsible for the growth of algae as well as the overall performance of MFCs such as light intensity, dissolved oxygen (DO), pH, carbon dioxide, etc. The light intensity surrounding the cathode chamber plays a key role in algal growth, which ultimately affects the oxygen generation and thereby the overall performance of A-MFCs [42]. In a recent study on A-MFC under several light/dark regimes, 18/6, 12/12, and 6/18h confirmed the strong dependence on illumination conditions on power production [43]. When the available light time was higher, i.e., 18/6 regime as compared to the other two conditions, a higher current production was observed in the A-MFC. The DO concentration was reduced to 0mg/L in the 6/18 light/dark cycle, which in turn resulted in no power generation into the MFC. Similarly, Wu et al. [21] found greater power density in a tubular photobioreactor MFC under continuous illumination as compared

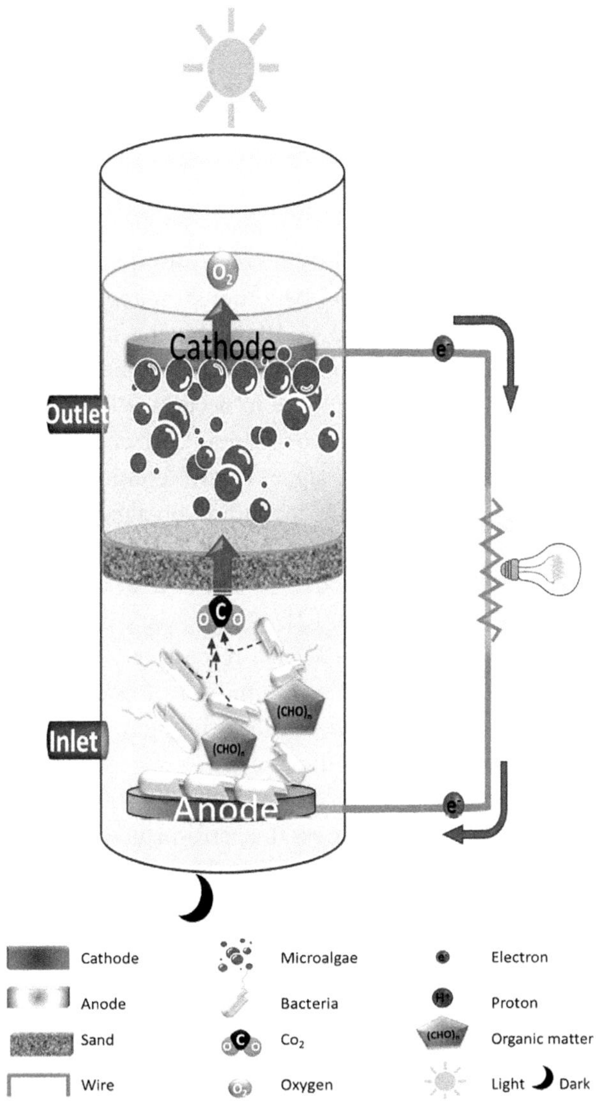

Fig. 3 Sediment algal MFC [25].

to intermittent illumination. A comparable study found that an A-MFC produced increased power density during higher light times, such as 24/0 as opposed to 16/8 and 12/12h [44]. According to the aforementioned data, algal photosynthesis actually took place in the cathode chamber, and the longer light time allowed for a higher power density in the A-MFC. It might be stated that during the day, algae could produce more oxygen through photosynthesis to support oxidation–reduction reactions (ORR);

however during the night, oxygen is used for respiration, causing DO depletion at the cathode chamber and lowering the overall performance of MFCs.

For instance, in the cathode chamber in MCCs, the dissolved oxygen concentration was consistently low (0.9 mg/L) [34]. However, likewise, it reduced consistently as the power generation increased. Thus, low dissolved oxygen concentrations are a good evident of maximum power output in PAMFCs using oxygen as the electron acceptor.

There are various studies in support of the impact of light on the performance of algae MFCs. The growth of algae is continuously increased in the light intensity range from 3500 lx–10,000 lx; thereafter, the PAMFC becomes saturated [44]. Photo-inhibition in algae is observed with light irradiance up to 10,000 lx, and it also occurs at a low light intensity range of 3500–5000 lx (light limitation). Photosynthesis is prohibited during the dark period, which typically results in little to no power output in PAMFCs. However, when light is again accessible, photosynthesis and energy production typically resume. To prevent photo-inhibition of algae, careful consideration must be given to choosing the ideal light intensity in MFCs depending on algae. Reduced power generation and algal growth have been observed when the light intensity is increased beyond saturation. PAMFCs are, therefore, made to work in a variety of light and dark conditions, such as 24/00, 16/8, and 12/12 h of light/dark conditions, with enhanced electricity output of 126, 112, and 72 mW/m^3, respectively [44]. A dark period must be maintained for algae respiration and consumption of oxygen. In a PAMFC with *C. vulgaris*, power generation was reduced when the light regime was reduced from 16/8 to 12/12 h (72%–57% coulombic efficiency).

Algal growth can be considerably improved by utilizing artificial lighting, such as light-emitting diodes (LEDs), which often improves the performance of MFCs [45]. When LED was employed as the light source instead of more traditional light sources such as compact fluorescent bulbs, the performance and efficiency of MFCs were improved [46]. When operated under both blue (0.156–4.741 mW/m^2) and red (1.525–12.947 mW/m^2) light, power density improves (from 100 to 900 lx) [46]. With the use of red LED light, it was found that the passage of electrons from the algal cell through the electrolyte to the electrode enhanced. This is because the photosynthetic system of algae is very efficient at absorbing light energy at this wavelength. Thus, the selection of a suitable light wavelength (white, red, yellow, blue, and green) is required in PAMFCs to achieve maximum power density [46]. For example, blue light affects photosynthesis (lower absorption of light energy), which in turn leads to lower power generation in PAMFCs [46]. PAMFCs are thus designed under different light wavelengths, for example, red LED produced a maximum power density of 12.947 mW/m^2. In a PAMFC with *Chlamydomonas reinhardtii*, power generation increased when the red LED light (12.947 mW/m^2) was replaced with blue LED light (4.741 mW/m^2).

Different species of algae have a substantial effect on the overall performance of A-MFCs as compared to other operational parameters. Among various algal strains,

C. vulgaris is frequently employed in the A-MFC due to its superior ability to fix CO_2 and oxygen generation through photosynthetic processes. Additional benefits of using *C. vulgaris* lie in the fact that they have a high concentration of fatty acids, which can be used to make biodiesel as a secondary fuel [44,47]. Hu et al. [48] carried out a study of two different algal species, viz. *C. vulgaris* and *C. autotrophica were* employed in a photosynthetic cathode chamber to check their potential for power density in air-lift-type microbial carbon capture cells (ALMCCs). 10.95% times higher power density was obtained from ALMCCs using *C. vulgaris* in the cathode. The difference in metabolism and growth parameters of these two algae resulted in different power outputs from A-MFC. The growth of *C. vulgaris* as compared to *C. autotrophica* at mentioned operational conditions was higher, thereby increasing the DO concentration in the cathode chamber while decreasing the internal resistance of the MFC. Furthermore, *C. vulgaris* has the ability to attach on the electrode and take part in the electron transport chain via some mediators. Therefore, it was assumed that the *C. vulgaris* attached to the cathode of MFC will be more efficient in power production as compared to the algae suspended in the cathode. Zhou et al. [49] cultured *C. vulgaris* in BG11 growth medium at 25–26°C temperature and 1600 lx cool-white fluorescent light intensity and immobilized as algae–alginate mixture. This immobilized algae mixture was then employed in an air-breathing A-MFC and the potential for power density was then compared with a dual-chamber A-MFC having suspended *C. vulgaris* in a cathode. Eighty-eight percent higher power density was observed with A-MFC having *C. vulgaris* immobilized cathode. The immobilization of *C. vulgaris* on the cathode in A-MFC has different advantages which include (i) algae act as terminal electron acceptor (TEA) (ii) low carbon dioxide mass diffusion resistance, and (iii) in situ production and utilization of oxygen at the biocathode which further reduces the internal resistance of A-MFC.

8 Applications of algae-mediated microbial fuel cells for wastewater treatment

Wastewaters generated from various processes are discharged to water bodies without any treatment, thus posing an adverse effect on the entire ecosystem health. Due to limited resources and freshwater on earth, wastewater should be exploited as a low-cost nutrient and water source. So, it is the need of the hour to necessarily treat the wastewater before discharging it into any water bodies. In the past few years, a paradigm shift has led to a new perception of wastewater, not as disruptive but as a source of energy and other valuable products. Considerable advances are required in the field of biotechnology to achieve environmental, economic, and energy sustainability by utilizing wastewater as - low-cost resource. MFC technology has been considered an efficient wastewater treatment device which additionally can generate useful products and bioelectricity. However, one of the major limitations linked with MFC operation for wastewater treatment is the limited nutrient removal by the bacteria present in the anode chamber. Oswald and Golueke [50] described the utilization of algae for wastewater remediation along with the potential of bioenergy generation from biomass. Algae's nutritional and chemical requirements can offer opportunities for advanced wastewater treatment by integrating waste/wastewater resources with algae systems for more holistic use [51]. Apart from bioelectricity generation, photosynthetic microorganisms (algae) can be applied to MFCs as wastewater treatment technology [52].

Depending on their sources, wastewaters can contain significant levels of nutrients (N and P), organic carbon, and trace elements. Algae are well renowned for having the ability to remove a variety of toxins from the domestic, industrial, and commercial growing sectors for wastewater pollution. It has been found that the survival rate of algae is much higher than other microorganisms in toxic wastewaters [53]. It has been well reported in the literature that mineralized forms of the various pollutants (ammonium (NH^{4+}), nitrate (NO^{3-}), phosphate (PO_4^{3-}), etc.) act as a nutrient source for the growth of these algae [30,43,54–56]. The nutrients present in the wastewater are used by algae for their biomass generation which, in turn, can be further processed for biofuel production or even using biomass for producing extra electricity, which makes this process more economical [53]. For biomass synthesis, many photoautotrophs, such as algae and cyanobacteria, have the ability to efficiently absorb nutrients from wastewater. The nitrogen removal by algae is due to assimilation while the phosphorus removal occurs by assimilation for algal growth and its uptake to be stored as polyphosphates [57]. The assimilation of nutrients is directly proportional to the growth of algae which means the higher the nutrient uptake, the higher the biomass concentration. The synergetic relationship between nutrient recovery and growth of algae in the cathode chamber of dual-chambered MFC is studied by various researchers [58]. The impact of closed and open circuits along with nutrient uptake was studied by Uggetti and Puigagut [54] on biomass concentration of algae. From the analysis results, it was observed that the closed circuit operation exhibited 1.5–3 times higher biomass concentration as well as 25% more phosphate removal than the open circuit mode of operation. However, the removal of ammonium was observed to be similar (~99%) with both the operations (open and closed circuit) in A-MFCs. The higher removal of ammonium by algae can be attributed to either its growth or its oxidation via nitrification. The significant increase in the concentration of nitrates is a strong evidence for nitrification during the operation, which can act as an efficient electron acceptor at the cathode [59]. Algae utilize phosphates in the form of phosphates precipitation at higher pH values. Zhang et al. [60] observed the 87.6% and 69.8%, removal of nitrogen and phosphorus by algae, respectively, in a sediment-type photomicrobial fuel cell (PFC) along with algae biomass yield of 0.56 kg/m^3. Additionally, enhanced removal efficiencies for nitrogen (98%) and phosphate (82%) were observed after optimization of the reactor conditions and buffering capacity [37].

The treatment of anaerobically digested effluent from kitchen waste (ADE-KW) was evaluated by Hou et al. [61] in both anode and cathode chambers of MFC at different dilutions. MFC combined with *Golenkinia* sp. SDEC-16 as a cathodic biocatalyst has shown ~90% of cathodic total nitrogen and total phosphorus removal efficiencies when operated with ADE-KW as substrate [61].

8.1 COD removal by algae-assisted MFC

Carbon is an important parameter required for the growth of algae. Based on the availability of carbon sources, algae are classified as photoautotrophic, heterotrophic, and mixotrophic mode of cultivation [55,62]. In the photoautotrophic mode of cultivation, algae utilize inorganic carbon (e.g., bicarbonates or CO_2) as the carbon source to produce carbohydrates through photosynthesis. However, in heterotrophic cultivation, organic carbon is utilized as a carbon source in the presence or absence of light. The composition of microbial community and bioenergy generation in MFCs is also influenced by the carbon type. The heterotrophic mode enhances power production, handling simplicity, and biofuel productivity as the heterotrophic growth of algae is often faster than autotrophic growth [58,62]. Colombo et al. [29] reported a similar removal efficiency for soluble chemical oxygen demand (sCOD (89%) in both the MFCs (anode chambers of A-MFC and air cathode MFC); however, the maximum power generation was achieved in the cathode chamber of A-MFC [44]. The enhanced removal efficiency for COD, other nutrients, and overall performance is observed by the use of algae in the cathode chamber of MFC even when the anodic effluent is fed to the cathode chamber. In past years, the efficiency of A-MFCs has been improved many folds for carbon removal and power production. [63]. Commault et al. [64] investigated the removal of chemical oxygen demand (COD) using *C. vulgaris* as the cathodic biocatalyst in a dual-chamber MFC from synthetic wastewater (anodic effluent). It has been found that extra 10%–15% of overall COD (49%) can be removed using algal biofilm in the cathode chamber with a retention time of 8 days. The lower efficiency of COD removal by *C. vulgaris* is attributed to its inability to remove dissolved organic carbon in the medium [64,65]. Movement of organics and nutrients from the anode to the cathode chamber of MFC leads to a decrease in COD removal at the cathode chamber. In the algae-based MFCs, different types of biomass can be obtained from the cathode chamber such as fresh and untreated algal biomass, algal biomass after lipid extraction, pretreated algal biomass, algal powder, and predigested algal biomass. To understand the feasibility of using algal biomass at different concentrations (1–5 g COD/L) in an anode chamber (as COD) for enhanced bioelectricity production has been reported by various researchers [66]. As the composition of raw algal biomass differs from pretreated algal biomass, a recent study reported the use of pretreated algal biomass of *Scenedesmus obliquus* in terms of total and soluble chemical oxygen demand to determine the applicability of algal biomass as the substrate in the cathode chamber of MFCs. The readily degradable pretreated algal biomass yielded maximum total and soluble COD removal efficiencies of 59.2% and 66.4%, respectively. Similarly, MFCs assisted with lipid-extracted algal biomass and fruit pulp waste exhibited a maximum COD removal efficiency of 70.8% and 62.3%, respectively [67].

8.2 Heavy metals removal potential of algae-assisted MFC

Nowadays, the high concentration of specific pollutants in wastewater of different industrial sectors is a major concern. To reduce the level of these pollutants in wastewater streams, various strict regulations have been imposed by the Water Framework Directive, REACH regulation, European Union IPPC, Industrial Emission Related Directive, etc. [55]. The wastewater generated from different industrial sectors such as electroplating, metal cleaning, etc., has a high concentration of heavy metals which is difficult to treat. However, algae is considered a potential candidate to remediate industrial wastewater loaded with heavy metals [51,68]. Various factors responsible for the removal of heavy metals from wastewater by algae-mediated MFCs are the type of algal strain, concentration, and type of heavy metals, pH, temperature, light, etc. [51,69]. The removal of heavy metals by algae includes two steps: first is adsorption into extracellular cell wall and second is adsorption inside the cells.

The removal efficiency of heavy metals by algae depends on the high affinity of metals to algae cell wall. The heavy metals remediation by algae is eased by class III metallothioneins (MtIII) [69]. Heavy metals having high redox potential can be used as electron acceptors in bioelectrochemical fuel cells, to get reduced and precipitated [13]. This process facilitates the recovery of maximum heavy metal ions from wastewater.

The maximum removal of Cd(II) up to 50 ppm concentration was observed by algae when nickel foam/graphene electrodes were used [70]. The maximum absorption of Cd(II) was observed to be 115 g/m^2 by *Chlorella* sp. *QB-102* in the cathode chamber of dual MFC. *Chlorella vulgaris* has been reported with maximum removal of Cd(II) by 86.6 mg/g by [71] in the cathode chamber of a batch stirred system. Therefore, these studies proved the better efficiency of algae-mediated MFC in a cathode chamber for the removal of heavy metals as well as power generation. A study carried out by Singhvi and Chhabra [72] observed 98% removal of Cr(VI) from wastewater using algal biomass in a dual-chamber salt bridge microbial fuel cell in 96 h along with a power density of 207 mWm^{-2}.

9 Biorefinery approach-based MFC systems for power generation and value-added chemical production

Various steps are involved in the production of value-added products from algae-based Microbial fuel cell. The first step involves harvesting the algal biomass (e.g., biofilm formation) from the cathode chamber. This can be achieved by using various low-cost harvesting techniques such as bioflocculation, cell immobilization, biofilm formation or membrane separation, etc. The next step deals with the recovery of value-added chemicals from microalgae which require pretreatment of algal biomass. Usually algal consortia are used in microbial fuel cells and their composition depends on the inoculum used. Five different types of pigments, chlorophyll, β-carotene, canthaxanthin, lutein, and zeaxanthin, are produced in a biorefinery process along with the dual advantage of power generation as well as with the dual role of generating electricity and value-added chemicals in MFCs. The algal biomass obtained from the biorefinery can be used as a source for biogas (CH_4) production, and the CO_2 produced in the process can be used as an inorganic carbon source for the growth of algae. Also, the algal biomass leftover after lipid extraction can be used as animal fodder. Genetic engineering and the use of algal biofilms proved to be efficient alternatives for desired algal biomass production along with compounds production. Greater technological advancement in the field of biofuel and value-added products generation can be achieved using genetically engineered algae species [73]. The high calorific value of algal biomass caters to significant advancement over other substrates used in biological processes and over conventional sources of energy [51]. Apart from biofuel and biomass production, the other uses of algal biomass are in colorant, medicines, cosmetics, and preservatives [74]. The algal species possess high concentration of lipids, which are suitable for biodiesel production [48].

10 Conclusion

Algae-mediated microbial fuel cells are pioneering bifunctional devices, which have a dual potential for wastewater treatment as well as renewable energy generation in a sustainable manner. The growth of algae is influenced by various operational parameters which, in turn, improves the overall performance of A-MFC. The various products formed from algal biomass facilitate technological advancement in the field of green energy and value-added products generation. The closed-loop biorefinery system along with various integrated processes in A-MFC can achieve zero waste discharge. Before using algae-mediated MFCs for wastewater treatment, further research is aimed at (i) the selection of suitable algal strains used in both the compartment (anode and cathode), (ii) control of cathodic reactions (electron transfer) and carbon dioxide uptake for enhanced power generation, biomass control, and biomass recovery, (iii) modeling of algae-based MFCs, and (iv) development cost-effective and sustainable technologies such as biorefinery approach with advantage of wastewater treatment biomass recovery and value-added products from the algal biomass.

References

[1] A. Bayrakdar, E. Tilahun, B. Çalli, Simultaneous nitrate and sulfide removal using a bio-electrochemical system, Bioelectrochemistry 129 (2019) 228–234.

[2] P. Kuntke, T.H.J.A. Sleutels, M. Rodríguez Arredondo, S. Georg, S.G. Barbosa, A. Ter Heijne, V.M. Hubertus, C.J.N. Buisman, (Bio) electrochemical ammonia recovery: progress and perspectives, Appl. Microbiol. Biotechnol. 102 (2018) 3865–3878.
[3] H. Yokoyama, H. Ohmori, M. Ishida, M. Waki, Y. Tanaka, Treatment of cow-waste slurry by a microbial fuel cell and the properties of the treated slurry as a liquid manure, Anim. Sci. J. 77 (6) (2006) 634–638.
[4] S. Wu, T. Lv, Q. Lu, Z. Ajmal, R. Dong, Treatment of anaerobic digestate supernatant in microbial fuel cell coupled constructed wetlands: evaluation of nitrogen removal, electricity generation, and bacterial community response, Sci. Total Environ. 580 (2017) 339–346.
[5] T. Kim, J. An, J.K. Jang, I.S. Chang, Coupling of anaerobic digester and microbial fuel cell for COD removal and ammonia recovery, Bioresour. Technol. 195 (2015) 217–222.
[6] T.P. Sciarria, S. Arioli, G. Gargari, D. Mora, F. Adani, Monitoring microbial communities' dynamics during the start-up of microbial fuel cells by high-throughput screening techniques, Biotechnol. Rep. 21 (2019) e00310.
[7] M. Cerrillo, J. Oliveras, M. Viñas, A. Bonmatí, Comparative assessment of raw and digested pig slurry treatment in bioelectrochemical systems, Bioelectrochemistry 110 (2016) 69–78.
[8] A. Sotres, M. Cerrillo, M. Viñas, A. Bonmatí, Nitrogen recovery from pig slurry in a two-chambered bioelectrochemical system, Bioresour. Technol. 194 (2015) 373–382.
[9] R.D. Cusick, B.E. Logan, Phosphate recovery as struvite within a single chamber microbial electrolysis cell, Bioresour. Technol. (2012) 110–115.
[10] Y. Wang, H. Zhang, Y. Feng, B. Li, M. Yu, X. Xu, L. Cai, Bio-Electron-Fenton (BEF) process driven by sediment microbial fuel cells (SMFCs) for antibiotics desorption and degradation, Biosens. Bioelectron. 136 (2019) 8–15.
[11] J.J. Fornero, M. Rosenbaum, M.A. Cotta, L.T. Angenent, Carbon dioxide addition to microbial fuel cell cathodes maintains sustainable catholyte pH and improves anolyte pH, alkalinity, and conductivity, Environ. Sci. Technol. 44 (7) (2010) 2728–2734.
[12] A.E. Inglesby, D.A. Beatty, A.C. Fisher, Rhodopseudomonas palustris purple bacteria fed Arthrospira maxima cyanobacteria: demonstration of application in microbial fuel cells, RSC Adv. 2 (11) (2012) 4829–4838.
[13] A.S. Mathuriya, J.V. Yakhmi, Microbial fuel cells to recover heavy metals, Environ. Chem. Lett. 12 (2014) 483–494.
[14] S.V. Mohan, G. Velvizhi, J.A. Modestra, S. Modestra, Microbial fuel cell: critical factors regulating biocatalyzed electrochemical process and recent advancements, Renew. Sust. Energ. Rev. 40 (2014) 779–797.
[15] B.E. Logan, B.E. Hamelers, R. Rozendal, U. Schröder, J. Keller, S. Freguia, P. Aelterman, W. Verstraete, K. Rabaey, Microbial fuel cells: methodology and technology, Environ. Sci. Technol. 40 (17) (2006) 5181–5192.
[16] S. Bajracharya, A. ElMekawy, S. Srikanth, D. Pant, Cathodes for microbial fuel cells, Microbial Electrochemical and Fuel Cells, Elsevier, 2016, pp. 179–213.
[17] S.K. Butti, G. Velvizhi, M.L. Sulonen, J.M. Haavisto, E.O. Koroglu, A.Y. Cetinkaya, S. Singh, D. Arya, J.A. Modestra, K.V. Krishna, A. Verma, B. Ozkaya, A.M. Lakaniemi, J.A. Puhakka, S.V. Mohan, Microbial electrochemical technologies with the perspective of harnessing bioenergy: maneuvering towards upscaling, Renew. Sust. Energ. Rev. 53 (2016) 462–476.
[18] D. Singh, D. Pratap, Y. Baranwal, B. Kumar, R.K. Chaudhary, Microbial fuel cells: a green technology for power generation, Ann. Biol. Res. 1 (3) (2010) 128–138.
[19] S. Freguia, E.H. Teh, N. Boon, K.M. Leung, J. Keller, K. Rabaey, Microbial fuel cells operating on mixed fatty acids, Bioresour. Technol. 101 (4) (2010) 1233–1238.
[20] P.T. Kelly, Z. He, Understanding the application niche of microbial fuel cells in a cheese wastewater treatment process, Bioresour. Technol. 157 (2014) 154–160.
[21] D. Wu,, D. Xing, X. Mei, B. Liu, C. Guo, N. Ren, Electricity generation by *Shewanella sp.* HN-41 in microbial fuel cells, Int. J. Hydrog. Energy 38 (35) (2013) 15568–15573.
[22] V.G. Gude, Wastewater treatment in microbial fuel cells—an overview, J. Clean. Prod. 122 (2016) 287–307.

[23] E.E. Powell, M.L. Mapiour, R.W. Evitts, G.A. Hill, Growth kinetics of Chlorella vulgaris and its use as a cathodic half cell, Bioresour. Technol. 100 (1) (2009) 269–274.
[24] D. Parlevliet, N.R. Moheimani, Efficient conversion of solar energy to biomass and electricity, Aquat. Biosyst. 10 (1) (2014) 4.
[25] M.E. Elshobary, H.M. Zabed, J. Yun, G. Zhang, X. Qi, Recent insights into microalgae-assisted microbial fuel cells for generating sustainable bioelectricity, Int. J. Hydrog. Energy 46 (4) (2020) 3135–3159.
[26] C.C. Lin, C.H. Wei, C.I. Chen, C.J. Shieh, Y.C. Liu, Characteristics of the photosynthesis microbial fuel cell with a Spirulina platensis biofilm, Bioresour. Technol. 135 (2013) 640–643.
[27] C.C. Fu, T.C. Hung, W.T. Wu, T.C. Wen, C.H. Su, Current and voltage responses in instant photosynthetic microbial cells with Spirulina platensis, Biochem. Eng. J. 52 (2–3) (2010) 175–180.
[28] K. Nishio, K. Hashimoto, K. Watanabe, Light/electricity conversion by defined cocultures of Chlamydomonas and Geobacter, J. Biosci. Bioeng. 115 (4) (2013) 412–417.
[29] A. Colombo, S. Marzorati, G. Lucchini, P. Cristiani, D. Pant, A. Schievano, Assisting cultivation of photosynthetic microorganisms by microbial fuel cells to enhance nutrients recovery from wastewater, Bioresour. Technol. 237 (2017) 240–248.
[30] C. Nagendranatha Reddy, R. Kakarla, B. Min, Algal biocathodes, in: S. Venkata Mohan (Ed.), Microbial Electrochemical Technology: Platform for Fuels, Chemicals and Remediation'. Biomass, Biofuels and Biochemicals, 2019, pp. 525–547.
[31] Y. Zhang, Y. Zhao, M. Zhou, A photosynthetic algal microbial fuel cell for treating swine wastewater, Environ. Sci. Pollut. Res. 26 (2019) 6182–6190.
[32] A. Schievano, A. Colombo, A. Cossettini, A. Goglio, V. D'Ardes, S. Trasatti, P. Cristiani, Single-chamber microbial fuel cells as on-line shock-sensors for volatile fatty acids in anaerobic digesters, Waste Manag. 71 (2018) 785–791.
[33] Q. Hou, C. Nie, H. Pei, W. Hu, L. Jiang, Z. Yang, The effect of algae species on the bioelectricity and biodiesel generation through open-air cathode microbial fuel cell with kitchen waste anaerobically digested effluent as substrate, Bioresour. Technol. 218 (2016) 902–908.
[34] X. Wang, Y. Feng, J. Liu, H. Lee, C. Li, N. Li, N. Ren, Sequestration of CO2 discharged from anode by algal cathode in microbial carbon capture cells (MCCs), Biosens. Bioelectron. 25 (12) (2010) 2639–2643.
[35] Z. Chi, J.V. O'Fallon, S. Chen, Bicarbonate produced from carbon capture for algae culture, Trends Biotechnol. 29 (11) (2011) 537–541.
[36] B.E. Logan, M.J. Wallack, K.Y. Kim, W. He, Y. Feng, P.E. Saikaly, Assessment of microbial fuel cell configurations and power densities, Environ. Sci. Technol. Lett. 2 (8) (2015) 206–214.
[37] L. Xiao, E.B. Young, J.A. Berges, Z. He, Integrated photo-bioelectrochemical system for contaminants removal and bioenergy production, Environ. Sci. Technol. 46 (2012) 11459–11466.
[38] L. De Schamphelaire, W. Verstraete, Revival of the biological sunlight-to-biogas energy conversion system, Biotechnol. Bioeng. 103 (2009) 296–304.
[39] H.J. Jeon, K.W. Seo, S.H. Lee, Y.H. Yang, R.S. Kumaran, S. Kim, H.J. Kim, Production of algal biomass (Chlorella vulgaris) using sediment microbial fuel cells, Bioresour. Technol. 109 (2012) 308–311.
[40] C.T. Wang, Y.S. Huang, T. Sangeetha, Y.M. Chen, W.T. Chong, H.C. Ong, W.M. Yan, Novel bufferless photosynthetic microbial fuel cell (PMFCs) for enhanced electrochemical performance, Bioresour. Technol. 255 (2018) 83–87.
[41] V.M. Luimstra, S.J. Kennedy, J. Güttler, S.A. Wood, D.E. Williams, M.A. Packer, A cost-effective microbial fuel cell to detect and select for photosynthetic electrogenic activity in algae and cyanobacteria, J. Appl. Phycol. 26 (1) (2014) 15–23.
[42] Y.C. Wu, Z.J. Wang, Y. Zheng, Y. Xiao, Z.H. Yang, F. Zhao, Light intensity affects the performance of photo microbial fuel cells with *Desmodesmus* sp. *A8* as cathodic microorganism, Appl. Energy 116 (2014) 86–90.
[43] R. Kakarla, B. Min, Sustainable electricity generation and ammonium removal by microbial fuel cell with a microalgae assisted cathode at various environmental conditions, Bioresour. Technol. 284 (2019) 161–167.

[44] E. Bazdar, R. Roshandel, S. Yaghmaei, M.M. Mardanpour, The effect of different light intensities and light/dark regimes on the performance of photosynthetic microalgae microbial fuel cell, Bioresour. Technol. 261 (2018) 350–360.
[45] J.C.W. Lan, K. Raman, C.M. Huang, C.M. Chang, The impact of monochromatic blue and red LED light upon performance of photo microbial fuel cells (PMFCs) using *Chlamydomonas reinhardtii* transformation F5 as biocatalyst, Biochem. Eng. J. 78 (2013) 39–43.
[46] H.B. Chen, J.Y. Wu, C.F. Wang, C.C. Fu, C.J. Shieh, C.I. Chen, et al., Modeling on chlorophyll a and phycocyanin production by *Spirulina platensis* under various light-emitting diodes, Biochem. Eng. J. 53 (1) (2010) 52–56.
[47] M. Khazraee Zamanpour, H.R. Kariminia, M. Vosoughi, Electricity generation, desalination and microalgae cultivation in a biocathode-microbial desalination cell, J. Environ. Chem. Eng. 5 (2017) 843–848.
[48] X. Hu, J. Zhou, B. Liu, Effect of algal species and light intensity on the performance of an air-lift-type microbial carbon capture cell with an algae-assisted cathode, RSC Adv. 6 (2016) 25094–25100.
[49] M. Zhou, H. He, T. Jin, H. Wang, Power generation enhancement in novel microbial carbon capture cells with immobilized *Chlorella vulgaris*, J. Power Sources 214 (2012) 216–219.
[50] W.J. Oswald, Golueke, C. G., Eutrophication trends in the United States: a problem? J. Water Pollut. Control Fed. (1966) 964–975.
[51] J.H. Hwang, J. Church, S.J. Lee, J. Park, W.H. Lee, Use of microalgae for advanced wastewater treatment and sustainable bio energy generation, Environ. Eng. Sci. (2016) 1–17.
[52] R.S. Berk, & Canfield, J. H., Bioelectrochemical energy conversion, Appl. Microbiol. 12 (1) (1964) 10–12.
[53] S. Luo, J.A. Berges, Z. He, E.B. Young, Algal-microbial community collaboration for energy recovery and nutrient remediation from wastewater in integrated photobioelectrochemical systems, Algal Res. 24 (2017) 527–539.
[54] E. Uggetti, J. Puigagut, Photosynthetic Membrane-Less Microbial Fuel Cells to enhance microalgal biomass concentration, Bioresour. Technol. 218 (2016) 1016–1020.
[55] F. Delrue, P.D. Alvarez-Diaz, S. Fon-Sing, G. Fleury, J.F. Sassi, The environmental bio refinery: using microalgae to remediate wastewater, awin-win paradigm, Energies 9 (2016) 2–19.
[56] C.N. Reddy, S.V. Mohan, Integrated bio-electrogenic process for bioelectricity production and cathodic nutrient recovery from azo dye wastewater, Renew. Energ. 34 (11) (2016) 852–855.
[57] N. Powell, A.N. Shilton, S. Pratt, Y. Chisti, Factors influencing luxury uptake of phosphorus by microalgae in waste stabilization ponds, Environ. Sci. Technol. 42 (2008) 5958–5962.
[58] Y. Cui, N. Rashid, N. Hu, M.S.U. Rehman, J.I. Han, Electricity generation and microalgae cultivation in microbial fuel cell using microalgae-enriched anode and bio-cathode, Energy Convers. Manag. 79 (2014) 674–680.
[59] C. Fang, B. Min, I. Angelidaki, Nitrate as an oxidant in the cathode chamber of a microbial fuel cell for both power generation and nutrient removal purposes, Appl. Biochem. Biotechnol. 164 (2011) 464–474.
[60] Y. Zhang, J.S. Noori, I. Angelidaki, Simultaneous organic carbon, nutrients removal and energy production in a photomicrobial fuel cell (PFC), Energy Environ. Sci. 4 (2011) 4340–4346.
[61] Q. Hou, J. Cheng, C. Nie, H. Pei, L. Jiang, L. Zhang, Z. Yang, Features of *Golenkinia* sp. and microbial fuel cells used for the treatment of anaerobically digested effluent from kitchen waste at different dilutions, Bioresour. Technol. 240 (2017) 130–136.
[62] O. Perez-Garcia, Y. Bashan, Microalgal heterotrophic and mixotrophic culturing for bio-refining: from metabolic routes to techno-economics, in: A. Prokop, R. Bajpai, M. Zappi (Eds.), Algal Bio Refineries, 2015, pp. 61–131, https://doi.org/10.1007/978-3-319-20200-6_3.
[63] J. Ma, Z. Wang, J. Zhang, T.D. Waite, Z. Wu, Cost-effective Chlorella biomass production from dilute wastewater using a novel photosynthetic microbial fuel cell (PMFC), Water Resour. 10 (2016) 356–364.
[64] A.S. Commault, O. Laczka, N. Siboni, B. Tamburic, J.R. Crosswell, J.R. Seymour, P.J. Ralph, Electricity and biomass production in a bacteria-*Chlorella* based microbial fuel cell treating wastewater, J. Power Sources 356 (2017) 1–11.

[65] R. Huarachi-Olivera, A. Duenas-Gonza, U. Yapo-Pari, P. Vega, M. Romero-Ugarte, J. Tapia, L. Molina, A. Lazarte-Rivera, D.G. Pacheco-Salazar, M. Esparza, Bioelectrogenesis with microbial fuel cells (MFCs) using the microalga *Chlorella vulgaris* and bacterial communities, Electron. J. Biotechnol. 31 (2018) 34–43.
[66] N. Rashid, Y.F. Cui, M.S.U. Rehman, J.I. Han, Enhanced electricity generation by using algae biomass and activated sludge in microbial fuel cell, Sci. Total Environ. 456 (2013) 91–94.
[67] A. Khandelwal, A. Vijay, A. Dixit, M. Chhabra, Microbial fuel cell powered by lipid extracted algae: a promising system for algal lipids and power generation, Bioresour. Technol. 247 (2018) 520–527.
[68] W.W. Li, H.Q. Yu, Z. He, Towards sustainable wastewater treatment by using microbial fuel cells-centered technologies, Environ. Eng. Sci. 7 (2014) 911–924.
[69] K. Suresh Kumar, H.U. Dahms, E.J. Won, J.S. Le, K.H. Shin, Microalgae- A promising tool for heavy metal remediation, Ecotoxicol. Environ. Saf. 113 (2015) 329–352.
[70] Y. Zhang, Q. He, L. Xia, Y. Li, S. Song, Algae cathode microbial fuel cells for cadmium removal with simultaneous electricity production using nickel foam/graphene electrode, Biochem. Eng. J. 138 (2018) 179–187.
[71] Z. Aksu, G. Donmez, Binary biosorption of cadmium (II) and nickel (II) onto dried *Chlorella vulgaris*: co-ion effect on mono-component isotherm parameters, Process Biochem. 41 (2006) 860–868.
[72] P. Singhvi, M. Chhabra, Simultaneous chromium removal and power generation using algal biomass in a dual chambered salt bridge microbial fuel cell, J. Bioremediat. Biodegrad. 4 (5) (2013).
[73] A. Bansal, O. Shinde, S. Sarkar, Industrial wastewater treatment using phycoremediation technologies and co-production of value-added products, Bioremediat. Biodegrad. 9 (2018) 1–10.
[74] Z. Baicha, M.J. Salar-Garcia, V.M. Ortiz-Martinez, F.J. Hernandez-Fernandez, A.P. de los Rios, N. Labjar, E. Lotfi, M. Elmahi, A critical review on microalgae as an alternative source for bioenergy production: a promising low cost substrate for microbial fuel cells, Fuel Process. Technol. 154 (2016) 104–116.

CHAPTER 4

Renewable sustainable bio-catalyzed electricity production: Challenges and prospects of algal-based bio-electrochemical systems

Kulvinder Bajwa[a], Sushil Kumar[b], Muhammad Yousuf Jat Baloch[c], Smita S. Kumar[d], and Narsi R. Bishnoi[e]

[a]University School for Graduates Studies, USGS, Chaudhary Devi Lal University, Sirsa, India
[b]Department of Physics, Chaudhary Devi Lal University, Sirsa, India
[c]College of New Energy and Environment, Jilin University, Changchun, China
[d]Department of Environmental Studies, Central University of Haryana, Jant-Pali, Mahendergarh, Haryana, India
[e]Department of Environmental Science and Engineering, Guru Jambheshwar University of Science and Technology, Hisar, India

1 Introduction

The world population is rapidly growing, and it is predicted to approach 9 billion people by 2050 [1]. Furthermore, as the economy grows, the need for energy resources increases, which leads to greater consumption of conventional fuels, as a result, more significant issues arise, such as an energy shortage and greater environmental deterioration. Future energy sources should be carbon neutral and renewable to reduce the environmental impact of greenhouse gases. The depletion of fossil fuels, as well as the technological limitations of biofuel technology, has prompted a quest for more sustainable energy setup technologies, such as bio-electrochemical systems (BES) [2,3]. Biomass items for energy generation, such as growing crops and plants, algae, and organic wastes, fall within this category. Both macroalgae and microalgae have piqued interest as potential resources for a bio-based economy from the perspective of bioenergy and food security [4]. Microalgae have been employed in high-value products and mass culture for biomass because they are very effective solar converters. At a broad scale, the cost of algae growth and harvesting of biomass, are limiting factors for microalgal technology. Microalgal film and high-rate algal ponds (HRAPs) [5] have become more visible in facilitating the recovery of nutrients such as phosphate and nitrate from wastewater [6] as well as the removal of toxic waste [7]. Microalgae utilize the end products (such as CO_2 and ammonia) of bacterial metabolism to provide aerobic bacteria with the oxygen they need to degrade organic molecules in this process. This technique can improve resource efficiency by converting eutrophication into a biomass-producing

Algae Based Bioelectrochemical Systems for Carbon Sequestration, Carbon Storage, Bioremediation and Bioproduct Generation
https://doi.org/10.1016/B978-0-323-91023-1.00001-6

opportunity. HRAPs for wastewater treatment could provide efficient tertiary wastewater treatment, be cost-effective too, and while also producing algal biomass for biofuel generation [8]. Algal biomass contains a high concentration of proteins, carbohydrates, lipids, and other macromolecules, making it suitable for processing into biofuel, charcoal, and other bio-products [9–11]. Nonetheless, due to energy-intensive procedures such as biomass processing, transportation to processing locations, and specialized lab facilities, algal biomass biofuel research has yet to be completely commercialized. Wet biomass can be employed in hydrothermal cracking, anaerobic digestion, and fermentation; however, due to the limitations mentioned above, they are not the most cost-effective solutions [12]. By avoiding shipping and particular lab treatments, local and effective exploitation of algal biomass is highly recommended for sustainable waste management. Furthermore, the effective utilization of algal biomass in direct energy manufacture could be a novel strategy for reducing algal blooms [13,14]. Microbial fuel cells (MFCs) are a potential renewable energy technology that uses photosynthetic organisms to generate direct electricity [15–17]. MFCs utilize microorganisms' anaerobic respiration to transform organic waste (fuel) into energy that may be used to power practical applications [18]. However, the MFCs' performance is still restricted, and simplifying the oxygen reduction reaction (for the cathodic half-cell) without using costly catalysts is one option to prolong the technology's life. Mathuriya and Yakhmi [19] stated that MFCs could also help with the bioremediation of particular contaminants and nutrients in wastewater. MFCs may also be used to recover heavy metals, decolorize dyes, and produce bioenergy such as bio-hydrogen, bio-methane, and even biomass [19,20]. As a result, MFCs provide both electricity production and wastewater treatment, making the process more environmentally friendly and economically feasible [21].

2 Microbial fuel cells' advantage and basic principle

Since the early 2000s, MFCs have been a hot topic of discussion. Microbial fuel cell (MFC) technology is gaining a lot of interest these days because of its ability to produce power and clean wastewater at the same time. It is because wastewater treatment necessitated a high-energy consumption of 20–45 kWh/person-year, resulting in a considerable quantity of greenhouse gas emissions [22]. Studies have been conducted employing MFCs with algal-based cathodes (AFCs) to generate energy and wastewater treatment to address the scale-up challenges caused by using only bacteria. Algae create O_2 during photosynthesis, which reduces operational costs by eliminating the need for sparging air. The biomass produced is also utilized as a biofuel with simultaneous wastewater treatment being both ecologically and economically beneficial [23].

Microbial fuel cell applications have recently been established in a variety of industrial areas without the need for intermediaries [24]. The kind of microorganism and other operational parameters, such as the type of bacteria and the type of MFCs technology, determine the type of new MFCs technology to be used. MFCs, for example, can be used as a supplemental ionic membrane during wastewater treatment to consume extra components that are consumed by water bacterium species [23]. MFC technology produces energy in a clean and repeatable manner. In addition, MFCs consume energy considerably more efficiently than traditional combustion cells. MFCs have a theoretical efficiency of more than 50% [24]. MFCs may get their nutrition from any organic substance. Microorganisms produce electricity flow and the major benefit of MFCs is that they oxidize organic materials without requiring any pressure, unlike conventional fuel cells. Cathodic and anodic bodies are found in these microbial fuel cells. The microorganisms oxidize the fuel, producing electrons and protons. Protons flow through the membrane and electrons are delivered to the cathodic body through an external connection. On the cathodic body, electrons and protons mix with oxygen to produce water. Several bacterium species having pili on their exterior membrane are capable of transferring electrons without the need for intermediaries. Considering the peculiarities of the MFCs technology as well as the operational traits, such as the types of bacteria, ionic membranes, and operating temperatures [24]. Bacteria with active electrochemical enzymes as cytochromes on their external membrane transmit electrons to external materials in direct MFCs [24].

2.1 Basic reactions in anodic and cathode chamber

A proton-exchange membrane divides the anode and cathode compartments in a conventional MFC. A continuous supply of oxygen for the reaction in the cathode is required in such a system, resulting in additional energy consumption for aeration. As a result, requiring continual aeration is clearly a restriction for MFC applications in the real world.

The basic reaction occurs in the anode compartment, where carbohydrates play an important role as a fuel, accompanied by microorganisms. A succession of anaerobe reactions takes place on the anode, resulting in electron generation [24]. There are two types of reactions in MFCs: glucose interactions with water and algal respiration responses as a green plant. The reaction of a hydrocarbon and water in the anode portion, according to the following equation, produces, under anaerobic conditions:

$$Cl_2H_{22}O_{11} + 13H_2O \rightarrow 12CO_2 + 48H^+ + 48e \quad (1)$$

The protons are then released into the water. The cathode part is through the membrane, while electrons are transported to the cathode portion via the electrode.

The electrolyte in the cathode is a salt solution.

Water steam is created using the following equation in the presence of air:

$$3/2O_2 + 6H + 6e^- \rightarrow 3H_2O \quad (2)$$

2.2 Algae as feedstock in MFCs

Among the different resources utilized as microorganisms, microalgae have been discovered to be a more power-sustainable substrate for biofuel, biodiesel, and power generation in a few decades due to its rapid growth, photosynthetic capability, cheap cost of environment incubation, and commercial feasibility. Microalgae other than *Microcystis aeruginosa* have been widely employed as electron donors or oxygen producers in photomicrobial fuel cells [25]. There have been very few studies that have reported MFCs' use of hazardous algal biomass [26–28]. For example, with the elimination of MC-LR, cyanobacteria were employed to power the single-chamber MFC [29]. In another work, the acidic fermented broth of cyanobacterial biomass was used in sediment microbial fuel cells [28].

Ma et al. [30] created a photosynthetic MFC by using wastewater for the production of *Chlorella* biomass and proved that the system was more sustainable from an energy and biomass energy point of view. Previous studies had certain intrinsic flaws, such as the production of acid fermentation broth, which may limit the use of MFC on a wide scale for the simultaneous degradation of HABs and MC-LR [28]. According to Li et al. [31], the impacts of internal resistance on the electron transmission mechanism of power generation were not thoroughly explored. To investigate alternate mechanisms for MFCs elimination by indigenous consortia, the microbial community participating in HAB biodegradation must be dissected [31].

Zhu et al. [32] studied the nitrification/denitrification capabilities of MFCs and discovered that MFCs significantly increased the removal of nitrogen efficiency. Salar-Garcia et al. [33] discovered that using catholyte from ceramic MFCs improved microalgae lysis and boosted power generation under light/dark cycle conditions. Likewise, Saba et al. [34] examined MFCs for wastewater treatment, energy generation, and biomass production as well as the impact of various factors on MFC energy output. Baicha et al. [35] explored the use of microalgae for MFC bioenergy production, as well as the utilization of CO_2 in the cathode chamber for the development of algal biomass. Other research, in addition to MFC, has looked at the usage of algae in bioelectrochemical systems and microbial desalination cells.

Saba et al. [36] evaluated the usage of $KFe(CN)_6$ as cathode integrated with microalgae *Nannochloropsis salina* for power production from MDCs and found that *Nannochloropsis salina* had the greatest desalination efficiency and $KFe(CN)_6$ had the highest power generation. Similarly, Xu et al. [37] examined several innovative MFC-integrated technologies, as well as their advancement, while also recommending a future study path. A comprehensive microalgae-based MFC system for bioelectricity production with power densities and electrodes is given in Table 1.

Table 1 MFC systems based on microalgae for bioelectricity generation.

Microalgae	Type of MFCs	Power generation	Electrodes		Source
Chlorella Vulgaris	Microbial fuel cells wastewater (MFCW)	Maximum power density of 24.4 mW/m^2	Anode: carbon felt	Cathode: carbon paper-coated Pt	Wu et al. [38]
Alga (*Majoutia,*)	Algal–Enzyme Microbial fuel Cell	Electricity flow equal to 0.4 V in about 160 min	Anode: copper	Cathode: aluminum	Otadi et al. [24]
(Gammaproteobacteria, Deltaproteobacteria, and Alphaproteobacteria) Infused with algal biomass/mixed-culture algae biomass	Multi-anode sediment microbial fuel cell (SMFC)	High-power density of 2965 mW/m^2	Anode: TCMM anodes	Cathode: platinum-coated titanium mesh	Taskan et al. [39]
Chlorella sorokiniana	Microbial carbon capture cell (MCC),	Power density 7.02 ± 0.09 W/m^3	Cathode Carbon felt (Panex_35 Zolex	Anode- carbon felt (Panex_35 Zolex	Neethu et al. [40]
Chlorella vulgaris	Algae-based microbial carbon capture cell (MCC)	Maximum power density of 187 mW/m^2	Anode: carbon fiber brushes	Cathode: carbon felt	Liu et al. [41]
Microcystis aeruginosa fed with algal biomass	Integrated microbial fuel cell	Maximum power density (83 mW/m^2)	Anode: graphite fiber brush	Cathode: graphite felt	Ali et al. [14]
Oscillatoria agardhii	Algal-photosynthetic nanostructured Microbial Fuel cell	Power density of 26.8 mWm^{-2}	Anode: nanotubes (MWCNTs)/MnO_2	Cathode: Ag/AgCl (KCl, sat.) as	Mahmoud et al. [42]
C. vulgaris (microalgae) and Ulva lactuca (macroalgae)	anodic single-chamber MFCs	Power density (980 mWm^{-2})	Anode: platinum	Cathode: platinum	Fu et al. [43]

Continued

Table 1 MFC systems based on microalgae for bioelectricity generation—cont'd

Microalgae	Type of MFCs	Power generation	Electrodes		Source
Scenedesmus sp, C. vulgaris	Dual-chamber Microbial Fuel Cell	The highest power density of 1900 mW m ^{2}	Anode: dead microalgae biomass	Cathode: CO_2 generated	Cui et al. [44]
Chlorella Vulgaris	Wetland microbial fuel cell	33.14 mW m^{-3} and	Anode: graphite granules	Cathode: carbon felt	Gupta et al. [45]
Scenedesmus obliquus	Double-chamber	153 mWm^{-2}	Anode: plain carbon paper	Cathode: platinum coated carbon paper	Kakarla and Min [46]
C. vulgaris	Double-chamber	13.5 mWm^{-2}	Anode: toray carbon cloths 1	Cathode: toray carbon cloths with 10% teflon	Del Campo et al. [15]
S. obliquus	Double-chamber	102 mWm^{-2}	Anode: toray carbon paper	Cathode: toray carbon paper	Kondaveeti et al. [47]
Arthrospira maxima	Double-chamber	20.5 mWm^{-2}	Anode: graphite	Cathode: graphite	(Inglesby et al. [48]
C. vulgaris	Double-chamber	62.7 mWm^{-2}	Anode: plain Graphite	Cathode: plain Graphite	Gouveia et al. [49]

2.3 Single-chambered microalgae fuel cells

Saba and his coworker reported [34] about bacterio-algal MFCs. The utilization of algae in MFCs has gained popularity owing to algae's capacity to photosynthesize ambient CO_2, generating both oxygen and biomass, so aiding the cathode process. The bacterio-algal MFC has several advantages, including (1) wastewater treatment, (2) electricity generation, (3) algal biomass cultivation and pigment synthesis, (4) CO_2 sequestration, and (5) production of oxygen. Researchers have created many forms of algal-based microbial fuel cells, as shown in Fig. 1. As per Logrono et al. [50], microalgal biocathodes were used in an air-exposed single-chamber microbial fuel cell. The reactors were tested for their ability to biodegrade real dye textile wastewater, while also producing bioelectricity.

Velasquez Orta et al. [51] examined bioelectricity production from a phytoplankton, *Chlorella vulgaris*, and a macrophyte, *Ulva lactuca*, in single-chamber microbial fuel cells (MFCs). MFCs were fed the two algae (in powder form), resulting in variations in energy recovery, degradation efficiency, and power densities. *Chlorella* provided higher energy per substrate mass (2.5 kWh/kg), while *U. lactuca* was entirely decomposed throughout a batch cycle of 73% COD. Khandelwal et al. [52] stated that a potential microbial fuel cell (MFC) system in which alga grows in the cathode chamber, algal biomass is collected, and lipids are extracted. The lipid-extracted algal biomass was then used as an electron donor substrate. The performance of LEA-fed MFCs was compared to that of fruit waste-fed MFCs with the LEA-fed MFC outperforming the FP-MFCs across the board. Lipid-extracted algal biomass-fed MFCs achieved a power density of $2.7\,Wm^{-3}$, which is 145% and 260%.

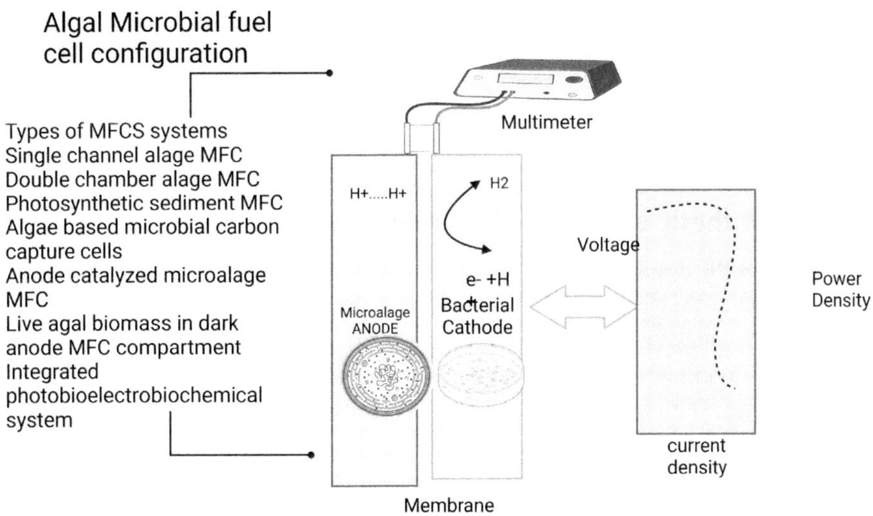

Fig. 1 Schematic representation of algal microbial fuel cell and its various types.

Shukla and Kumar [53] discuss the process of transfer of electrons in algae and algae to electrode, as well as the parameters influencing PAMFC performance. Arun et al. [54] investigated the use of algae-based microbial fuel cells for bioelectricity production, with a focus on the use of bacteria in the cathodic compartment, microalgae in the anodic compartment, and the major interactions between the compartments impacting bioelectricity production.

2.4 Double-chambered microbial fuel cells

In general, double-chambered fuel cells have an anodic chamber and a cathodic chamber separated by a PEM that permits proton transfer from the anode to the cathode while preventing oxygen access into the anode. As a result, this arrangement is often utilized to clean wastewater while also generating power. Both the cathode and anode are distinct compartments that are linked to one another via a proton-exchange membrane (PEM) [55]. In a dual-chamber MFC, immobilized anaerobic bacteria and cathodic *Chlorella* sp. generated electricity in the cathode chamber while treating swine dung. The effluents from photo-fermentation, anaerobic hydrogen, or methane fermentation systems may be used using microalgae-based MFC technology. In these systems, microalgae biomass may be used to generate either bio-hydrogen or bio-methane or sold as an additional source of revenue to offset the costs of MFC. Using this technology for both power production and wastewater treatment proved to be a success. To make bio-electrochemical systems more economically viable, a bio-refinery system is necessary [56].

Liu et al. [41] created a two-chambered MFC with a photosynthetic microalgae cathode that is driven by taking CO_2 from the anode and solar energy. Under light illumination, the anode off-gas is fed into the catholyte, resulting in a maximum power density of 187 mW/m^2, which is greater than the dark value of 21 mW/m2, indicating the critical role of algal photosynthesis. Cui et al. [44], Kondaveeti et al. [47], and Rashid et al. [57] studied the MFC by growing live green algae in the cathodic chamber and dead microalgal biomass as anodic biofilm substrate.

2.5 Photosynthetic sediment algal MFC

Biological anodes and abiotic cathodes are generally separated by a proton-exchange membrane in microbial fuel cells (MFCs). Typically, the abiotic cathode uses expensive catalysts such as platinum to catalyze the reduction of oxygen to form water [58]. As reported by Zabihallahpoor et al. [59] in MFCs, the cathodic reaction is frequently limiting, and researchers are currently concentrating on efficient, low-cost catalysts to enhance oxygen reduction at the cathode. Sediment MFCs are regarded as promising technologies for powering various sensors and electronics working in remote sensing locations. In the case of MFCs, exoelectrogenic bacteria in an anode chamber build a biofilm on the anode and transfer electrons extracellularly to it, either directly by using cytochromic membrane or indirectly (by conductive pili) and by using soluble endogenous mediators [59].

In a sediment-type MFC without a proton-exchange membrane, Commault and his colleagues [60] built a photosynthetic biocathode. There was no catalyst on the carbon and stainless steel cathode, but it was coated with a biofilm made up of a diverse population of microalgae and cyanobacteria [60]. As per the previous study conducted by Zhao et al. [28], the possibility of producing power from cyanobacteria in Taihu Lake using sediment MFC was examined, and its performance was compared to that of glucose and acetate-fed SMFCs. An SMFC powered by acidic fermentation broth obtained from Taihu Lake cyanobacteria produced 72 mWm2, had a COD removal rate of 76.2%, and a substrate degradation rate of 76.2%.

Song et al. [61] proposed that sediment microbial fuel cells have been limited in their use due to a shortage of electron acceptors in the cathode (SMFCs). *C. vulgaris* was used to generate oxygen as an electron acceptor in this experiment. The synergistic effects of *C. vulgaris* and electrogenic microorganisms in SMFCs were examined, and it was discovered that they improved organic matter biodegradation in sediments and converted chemical energy to electrical energy.

2.6 Algae-based microbial carbon captured microbial fuel cells

A new device known as a microbial carbon capture cell (MCC) was developed to reduce system CO_2 emissions by simply directing anodic off-gas into an MFCs' algal cathode chamber, as shown in Fig. 2, allowing for simultaneous wastewater treatment and electricity generation without the use of any energy [62]. By pumping anodic off-gas into an algae-growing cathode, new microbial carbon capture cells were created and

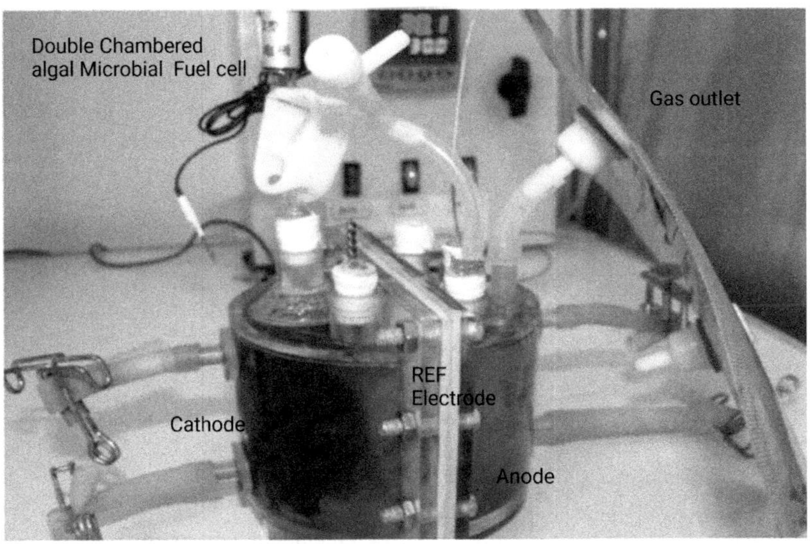

Fig. 2 Schematic presentation of two-chambered microbial carbon captured fuel cell. *(Modified A.H. Alami, S. Alasad, M. Ali, M. Alshamsi, Investigating algae for CO2 capture and accumulation and simultaneous production of biomass for biodiesel production, Sci. Total Environ. 759 (2021) 143529.)*

demonstrated to be a successful technology for CO_2 emission reduction with concomitant voltage output without aeration (*C. vulgaris*) [62].

Alami et al. [63] published a thorough and in-depth paper on the use of algae for carbon capture and storage. This is done in combination with algae cultivation for biomass generation for biodiesel manufacturing. To successfully decrease carbon emissions, many technologies for algae cultivation and carbon capture are being researched. Gupta et al. [45] developed a three-stage design of an algal-aided built wetland microbial fuel cell (Fig. 3) integrated sand filter for bioelectricity production and effective wastewater treatment was carried out by simulating a core of it.

Another recent work published by Neethu et al. [40] used response surface approach to optimize microalgae growth parameters while concurrently generating electricity in a carbon capture microbial fuel cell. Certain important elements influence the performance of a microbial carbon capture cell (MCC), which combines the benefits of a microbial fuel cell with a photo-bioreactor.

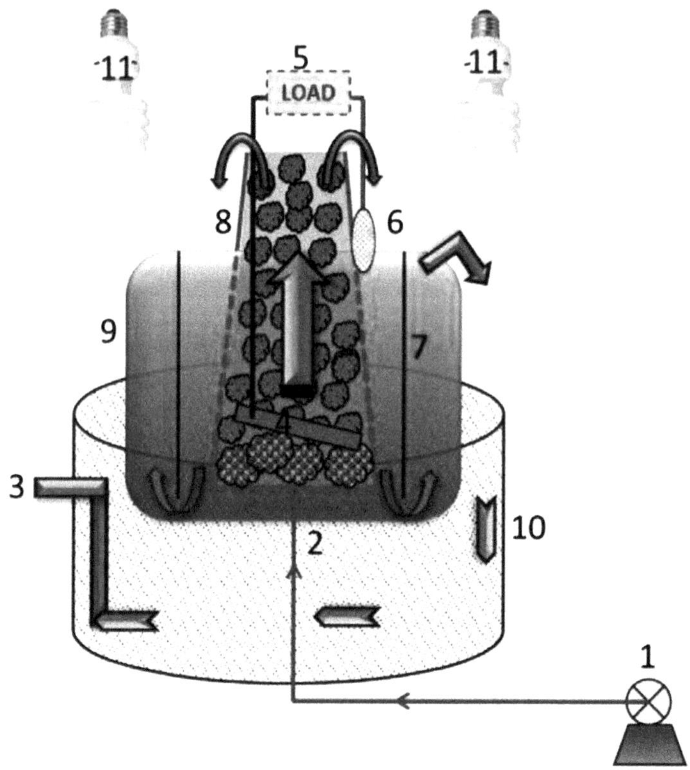

Fig. 3 Schematic representation of experimental setup of the constructed wetland cum microbial fuel cell with integrated sand chamber (1) peristaltic pump, (2) Influent line, (3) Effluent port, (4) graphite rod charge collector from the anode, (5) resistance, (6) carbon felt cathode, (7) baffle, (8) anode chamber (earthen pot), (9) cathode chamber (algal pond), (10) sand chamber, (11) light source, (/) water flow direction [45].

2.7 Anode-catalyzed microalgae microbial fuel cells (MFC)

One of the most critical components of an MFC is the anode, which offers the essential surface area for bacterial growth. These bacteria also produce electrons and protons, which they transmit to the anode. However, developing anode materials for MFC performance remains a problem [64]. Elshobary et al. [65] stated that the use of these photosynthetic organisms (algae) in MFC could improve efficiency and provide a cost-effective and renewable method for bioelectricity generation as shown in Fig. 4. Microalgae are frequently used in MFCs' anode and cathode compartments to produce electrons and oxygen, respectively.

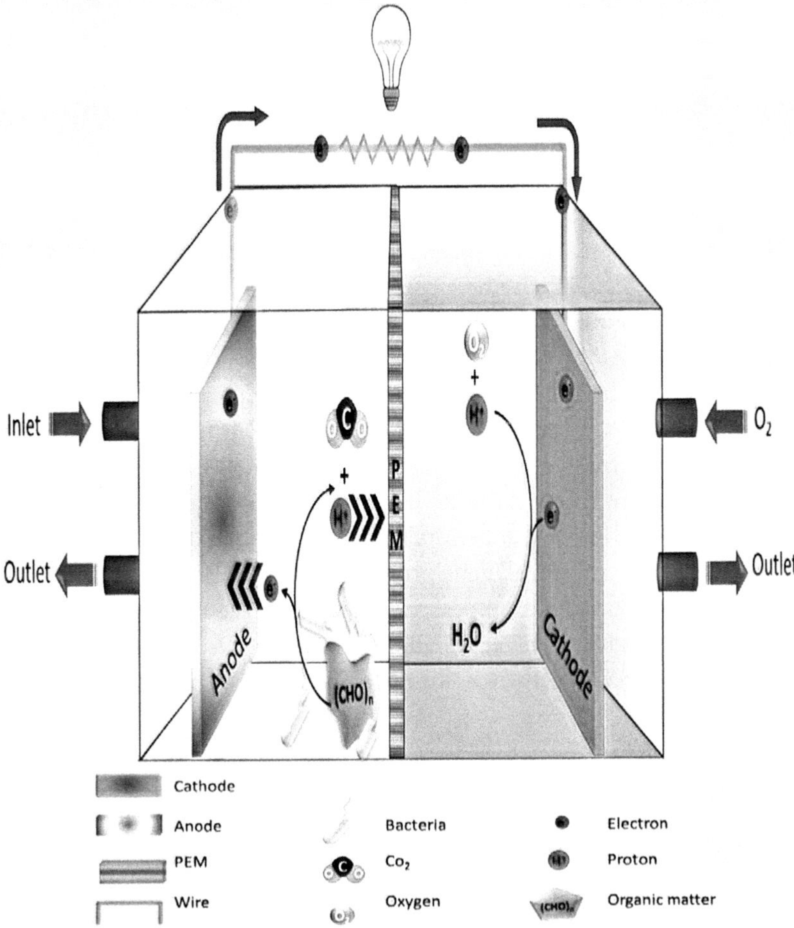

Fig. 4 Schematic diagram of a typical microbial fuel cell with anode chamber and cathode chamber sandwiched with a proton-exchange membrane (PEM). *(Source: M.E. Elshobary, H.M. Zabed, J. Yun, G. Zhang, X. Qi, Recent insights into microalgae-assisted microbial fuel cells for generating sustainable bioelectricity, Int. J. Hydrog. Energy 46(4) (2021) 3135–3159.)*

Likewise, Mekuto et al. [27] mentioned that microalgae-assisted MFC is a self-sustaining microbial fuel cell that uses a microalgae-assisted cathode to aid oxidation/reduction processes (ORR) while recycling the produced algal biomass to the anode compartment as a feedstock for increased energy output.

2.8 Live algal biomass in dark anode MFC compartment

Velasquez Orta et al. [51] constructed live algal biomass-based MFC by using *Ulva lactuca* and *Chlorella vulgaris* dry powder feedstocks in the anodic chamber of the system, and discovered that *C. vulgaris* had a maximum power density ($0.98\,Wm^{-2}$) and *U. lactuca* had a maximum power density of $760\,mW\,m^{-2}$. Rashid et al. [57] using microalgae *Scenedesmus* biomass as anode nutrition sources, activated sludge discovered that sonication and heat pretreatment of algal biomass increased the algae's microbial digestibility and the overall performance of the MFC test unit. A recent study conducted by Aiyer et al. [66] found that microbial communities function as catalysts in the operation of microbial fuel cells (MFCs). The utilization of a specified coculture of *Pseudomonas aeruginosa* and *Escherichia coli* enhanced power generation in MFCs. The coculture produced a higher power density of $190.44\,mWm^2$ *C. vulgaris*, a photosynthetic alga when introduced in a cathodic chamber. The mean power density increased by 41.7% in cocultures with *C. vulgaris*, from 175 to $248\,mW\,m^2$. When the cocultures were combined with *C. vulgaris*, a synergistic effect was found. MFCs with cocultures have many potential benefits for investigating processes and broadening the scope of applications [66].

Taskan et al. [39] evaluated the energy generation capabilities of mixed culture algal biomass using a multianode sediment microbial fuel cell in one of their latest studies (SMFC). A multianode reactor was used, with a tin-coated copper mesh anode and a platinum-coated titanium mesh cathode. The SMFC produced a high power density of $2965\,mW/m^2$, the highest power density reported in SMFC research to date. According to microscopic observations and gene profile investigations, TCCM's biocompatibility improves bacterial adhesion and enrichment of electroactive bacterial groups (*Gammaproteobacteria, Deltaproteobacteria*, and *Alphaproteobacteria*). Algal biomass could be used as an appropriate feedstock in SMFC to significantly improve electricity generation [39].

2.9 Integrated photo-biochemical microbial fuel cell system

By integrating an MFC into an algae-based bioreactor, several researchers have built integrated photobio-electrochemical systems. It has been discovered that such a device may be used to generate power and algal biomass. COD removal of approximately 92%, NH_4N (ammonium nitrogen removal) of 98%, and PO_4^- removal efficiency of 82% was reported by Xiao et al. [67] with a high power density output of $2.2\,Wm^{-3}$. Sun et al. [68] created a unique PBR integrated with stacked horizontal planar waveguide modules (SHPW-PBR) to mitigate the unfavorable effects of uneven spatial light distributions within associated cultivation photo-bioreactors on linked microalgae biofilm development (PBRs). Unlike traditional PBRs, the emerging light from external LED light bars was directed and uniformly disseminated within the SHPW-PBR by planar waveguides, providing light energy for the photoautotrophic development of microalgae cells. The average light intensity lighting the connected PBR is lower than the control PBR. *C. vulgaris* biofilm production increased by 204.11% in the SHPW-PBR, contributing to a 145.20% increase in areal *C. vulgaris* biofilm production. Following that, the responses of attached *C. vulgaris* biofilm development in the SHPW-PBR to various light intensities were assessed, and after 9 days of cultivation, the maximum areal *C. vulgaris* biofilm density reached 90.43 $g\,m^2$ at a light intensity of 136 $mol\,m^2\,s^1$ [68].

3 Scalability of MFC

The performance of MFC is governed by numerous elements such as system materials, architecture, microbe population, substrate supply, and system operating conditions [56]. The MFCs' scalability is determined not only by their fuel efficiency and capacity but also by the amount of electrical power they can create. The most difficult aspect of scaling up an MFC system is achieving enough energy output while utilizing an acceptable amount of reactor liquid. MFCs can be commercialized; bioreactors and the processes that go with them must be effectively scaled up to a big scale, not just a pilot size. These need considerably larger bioreactors. However, when the capacity of an MFC rises, the volumetric power production falls proportionately (power/unit volume). For scale-up, several miniatures of MFC stacks would be used. Recently, the modularization of many tiny MFCs and subsequent stacking for scale-up to reduce power or voltage loss has been established. Regarding the estimation of industrial cost for electricity generation, costing is a major aspect of scaling up this process.

4 Future perspectives

MFC technology is being promoted because it decreases the environmental carbon footprint, assures power supply, and provides efficient wastewater treatment all at the same time. Replacing pricey catalysts with microalgae has the potential to lower MFC production costs and treat wastewater, which would assist the wastewater treatment and power industries. Microalgae consume nutrients, which results in effective COD elimination and assists in the treatment of severe soil and water pollution [56]. Researchers observed filamentous algae, sometimes known as "macroalgae." In terms of not just nutrient uptake from wastewater but also harvesting, macroalgae had various benefits over microalgae [69–72]. If innovative reactor designs to fulfill the vast areal needs are constructed, filamentous algae's multiple benefits [73] can be used. Given the lack of study in this area, it is worth looking into the possibility of employing filamentous algae in MFCs. Future research in algal-based MFC technology must address present challenges from a technical standpoint while also considering the possibilities of altering algae cellular biology. This would lead to more advanced technology that could not only generate more electricity but also help with wastewater treatment [72,74].

For efficient algal growth, a variety of photo-bioreactor designs are available (Duan and Shi [75]). ,Innovative approaches to convert or combine these into MFCs can aid in determining the most effective configuration for growing algae in AFCs. The reactor design is influenced by elements such as the organism being employed, the anolyte and catholyte used, the available light source, and cost concerns. As a result, there is no set design for growing a certain algae species in a fuel cell. To summarize, future studies might focus on the following areas [72]:

- How an AFC's (algal-based fuel cell system) reactor components affect its power generation.
- Analyzing the influence of reactor components on the power generated in an AFC.
- AFC system mathematical modeling.
- Taking into consideration both the performance and economics of the AMFC system's scale-up.
- Using genetic engineering principles to increase electricity production.
- Inventive thinking.

5 Concluding remarks

Due to growing energy demand and a scarcity of present energy supplies, we must seek new energy sources that are clean, renewable, highly efficient, and cost-effective. Microbial fuel cell is one of the innovative energy-generating features that dependably meet the aforementioned requirements. Traditional energy sources are scarce and take a long time to regenerate. They are also inefficient, costly, and harmful to the environment. Microalgae-based MFC combines many tasks while producing electricity. Microalgae-based MFC uses bacteria and microalgae to create energy via photosynthesis and metabolic reaction while treating wastewater with high nutrient loading (necessary for microalgae growth) [56]. The three primary types of MFC reactor designs employing algae that have been studied (PSMFC) are microbial carbon capture cells based on algae, anode-catalyzed microalgal MFCs, and algal biomass or live algae as a substrate in the MFCs' dark anode chamber, and integrated photobio-electrochemical systems are the algal-based single- and double-chambered MFC, and photosynthetic sediment MFC [76]. Most MFC systems have their own set of benefits and drawbacks, but they all deliver the needed energy output patterns and working efficiency to varied degrees of complexity and power intensity outputs. Within the MFCs, none of these flaws can be found. Despite the fact that MFCs are used as suitable energy sources in various industrial and pilot operations, the generated power flow is too low. As a result, they are primarily found in sensors.

References

[1] D.H. Lee, Econometric assessment of bioenergy development, Int. J. Hydrog. Energy 42 (2017) 27701–27717.
[2] T.D. Searchinger, T. Beringer, A. Strong, Does the world have low-carbon bioenergy potential from the dedicated use of land? Energy Policy 110 (2017) 434–446.
[3] G. Joshi, J.K. Pandey, S. Rana, D.S. Rawat, Challenges and opportunities for the application of biofuel, Renew. Sust. Energ. Rev. 79 (2017) 850–866.
[4] C.S. Jones, S.P. Mayfield, Algae biofuels: versatility for the future of bioenergy, Curr. Opin. Biotechnol. 23 (3) (2012) 346–351.
[5] R.J. Craggs, S. Heubeck, T.J. Lundquist, J.R. Benemann, Algal biofuels from wastewater treatment high rate algal ponds, Environ. Sci. Technol. 63 (2011) 660e665.
[6] I. Di Termini, A. Prassone, C. Cattaneo, M. Rovatti, On the nitrogen and phosphorus removal in algal photobioreactors, Ecol. Eng. 37 (2011) 976e980.
[7] R. Munoz, B. Guieysse, Algal-bacterial processes for the treatment of hazardous contaminants: a review, Water Res. 40 (2006) 2799–2815.
[8] J.B.K. Park, R.J. Craggs, A.N. Shilton, Wastewater treatment high rate algal ponds for biofuel production, Bioresour. Technol. 102,35e42 (2011).
[9] M.S. Alfarisi, Sustainable bioethanol production from microalgae through ionic liquid as a potential catalyst, in: AIP Conference Proceedings. AIP Publishing LLC, 2020, p. 050003.
[10] J. Greenman, I. Gajda, I. Ieropoulos, Microbial fuel cells (MFC) and microalgae; photo microbial fuel cell (PMFC) as complete recycling machines. Sustainable, Energy Fuel 3 (2019) 2546e2560.
[11] S.V. Mohan, G. Nikhil, P. Chiranjeevi, C.N. Reddy, M. Rohit, A.N. Kumar, O. Sarkar, Waste biorefinery models towards sustainable circular bioeconomy: critical review and future perspectives, Bioresour. Technol. 215 (2016) 2–12.
[12] X. Yuan, X. Shi, D. Zhang, Y. Qiu, R. Guo, L. Wang, Biogas production and *microcystin* biodegradation in anaerobic digestion of blue algae, Energy Environ. Sci. 4 (2011) 1511e1515.
[13] J.D. Allan, S.D. Smith, P.B. McIntyre, C.A. Joseph, C.E. Dickinson, A.L. Marino, R.G. Biel, J.C. Olson, P.J. Doran, E.S. Rutherford, Using cultural ecosystem services to inform restoration priorities in the Laurentian Great Lakes, Front. Ecol. Environ. 13 (2015) 418e424.
[14] J. Ali, L. Wang, H. Waseem, B. Song, R. Djellabi, G. Pan, Turning harmful algal biomass to electricity by microbial fuel cell: a sustainable approach for waste management, Environ. Pollut. 266 (2020) 115373.
[15] A.G. del Campo, P. Cañizares, M.A. Rodrigo, F.J. Fernández, J. Lobato, Microbial fuel cell with an algae-assisted cathode: a preliminary assessment, J. Power Sources 242 (2013) 638–645.
[16] D.P. Strik, H.V. Hamelers, C.J. Buisman, Solar energy powered microbial fuel cell with a reversible bioelectrode, Environ. Sci. Technol. 44 (1) (2010) 532–537.
[17] M.C. Potter, Electrical effects accompanying the decomposition of organic compounds, Proc. R. Soc. B: Biol. Sci. 84 (571) (1911) 260–276.
[18] I.A. Ieropoulos, P. Ledezma, A. Stinchcombe, G. Papaharalabos, C. Melhuish, J. Greenman, Waste to real energy: the first MFC powered mobile phone, Phys. Chem. Chem. Phys. 15 (37) (2013) 15312–15316.
[19] A.S. Mathuriya, J.V. Yakhmi, Microbial fuel cells–applications for generation of 599 electrical power and beyond, Crit. Rev. Microbiol. 42 (1) (2016) 127–143.
[20] S.V. Mohan, G. Velvizhi, K.V. Krishna, M.L. Babu, Microbial catalyzed 606 electrochemical systems: a bio−factory with multi−facet applications, Bioresources 607 Technol. 165 (2014) 355–364.
[21] B.E. Logan, B. Hamelers, R. Rozendal, U. Schröder, J. Keller, S. Freguia, P. Aelterman, W. Verstraete, K. Rabaey, Microbial fuel cells: methodology and technology, Environ. Sci. Technol. 40 (2006) 5181–5192.
[22] S. Bolognesi, D. Cecconet, A. Callegari, A.G. Capodaglio, Combined microalgal photobioreactor/ microbial fuel cell system: performance analysis under different process conditions, Environ. Res. 192 (2021) 110263.
[23] V. Vinayak, M.J. Khan, S. Varjani, G.D. Saratale, R.G. Saratale, S.K. Bhatia, Microbial fuel cells for remediation of environmental pollutants and value addition: special focus on coupling diatom microbial fuel cells with photocatalytic and photoelectric fuel cells, J. Biotechnol. 338 (2021) 5–19.

[24] M. Otadi, S. Poormohamadian, F. Zabihi, M. Goharrokhi, Microbial fuel cell production with alga, World Appl. Sci. J. 14 (91) (2011).

[25] F. Ndayisenga, Z. Yu, Y. Yu, C.-H. Lay, D. Zhou, Bioelectricity generation using microalgal biomass as electron donor in a bio-anode microbial fuel cell, Bioresour. Technol. 270 (2018) 286e293.

[26] J. Ali, A. Sohail, L. Wang, M.R. Haider, S. Mulk, G. Pan, Electro-microbiology as a promising approach towards renewable energy and environmental sustainability, Energies 11 (2018) 1e30.

[27] L. Mekuto, A.V. Olowolafe, S. Pandit, N. Dyantyi, P. Nomngongo, R. Huberts, Microalgae as a biocathode and feedstock in anode chamber for a self-sustainable microbial fuel cell technology: a review, S. Afr. J. Chem. Eng. 31 (2020) 7e16.

[28] J. Zhao, X.F. Li, Y.P. Ren, X.H. Wang, C. Jian, Electricity generation from Taihu Lake cyanobacteria by sediment microbial fuel cells, J. Chem. Technol. Biotechnol. 87 (2012) 1567–1573.

[29] Y. Yuan, Q. Chen, S. Zhou, L. Zhuang, P. Hu, Bioelectricity generation and *microcystins* removal in a blue-green algae powered microbial fuel cell, J. Hazard. Mater. 187 (2011) 591–595.

[30] J. Ma, Z. Wang, J. Zhang, et al., Cost−effective *Chlorella* biomass production from dilute wastewater using a novel photosynthetic microbial fuel cell (PMFC), Water Res. 595 (108) (2017) 356–364.

[31] H. Li, H. Ai, L. Kang, X. Sun, Q. He, Simultaneous Microcystis algicidal and *microcystin* degrading capability by a single Acinetobacter bacterial strain, Environ. Sci. Technol. 50 (2016) 11903–11911.

[32] G. Zhu, G. Chen, R. Yu, et al., Enhanced simultaneous nitrification/denitrification in the biocathode of a microbial fuel cell fed with cyanobacteria solution. Process 722, Biochemist 51 (1) (2016) 80–88.

[33] M.J. Salar-García, I. Gajda, V.M. Ortiz-Martínez, et al., Microalgae as substrate in low cost terracotta-based microbial fuel cells: novel application of the catholyte produced, Bioresour. Technol. 209 (2016) 380–385.

[34] B. Saba, A.D. Christy, Z. Yu, A.C. Co, Sustainable power generation from bacterio-algal microbial fuel cells (MFCs): an overview, Renew. Sust. Energ. Rev. 73 (2017) 75–84.

[35] Z. Baicha, M.J. Salar-García, V.M. Ortiz-Martínez, et al., A critical review on microalgae as an alternative source for bioenergy production: a promising low cost substrate for microbial fuel cells, Fuel Process. Technol. 154 (2016) 104–116.

[36] B. Saba, A.D. Christy, Z. Yu, et al., Simultaneous power generation and desalination of microbial desalination cells using *Nannochloropsis salina* (marine algae) versus potassium ferricyanide as catholytes, Environ. Eng. Sci. (2017), https://doi.org/10.1089/ees.2016.0291.

[37] L. Xu, Y. Zhao, L. Doherty, et al., The integrated processes for wastewater treatment based on the principle of microbial fuel cells: a review, Crit. Rev. Environ. Sci. Technol. 46 (1) (2016) 60–91.

[38] X.Y. Wu, T.S. Song, X.J. Zhu, P. Wei, C.C. Zhou, Construction and operation of microbial fuel cell with *Chlorella vulgaris* biocathode for electricity generation, Appl. Biochem. Biotechnol. 171 (8) (2013) 2082–2092.

[39] B. Taşkan, M. Bakır, E. Taşkan, Enhanced power generation from algal biomass using multi-anode membrane-less sediment microbial fuel cell, Int. J. Energy Res. 45 (2) (2020) 2011–2022.

[40] B. Neethu, V. Tholia, M.M. Ghangrekar, Optimizing performance of a microbial carbon-capture cell using box-Behnken design, Process Biochem. 95 (2020) 99–107.

[41] Y. Liu, F. Harnisch, K. Fricke, R. Sietmann, U. Schröder, Improvement of the anodic bioelectrocatalytic activity of mixed culture biofilms by a simple consecutive electrochemical selection procedure, Biosens. Bioelectron. 24 (4) (2008) 1006–1011.

[42] R.H. Mahmoud, S.M. Abdo, F.A. Samhan, M.K. Ibrahim, G.H. Ali, R.Y. Hassan, Biosensing of algal-photosynthetic productivity using nanostructured bio-electrochemical systems, J. Chem. Technol. Biotechnol. 95 (4) (2020) 1028–1037.

[43] C.C. Fu, T.C. Hung, W.T. Wu, T.C. Wen, C.H. Su, Current and voltage responses in instant photosynthetic microbial cells with *Spirulina platensis*, Biochem. Eng. J. 52 (2–3) (2010) 175–180.

[44] Y.F. Cui, N. Rashid, N.X. Hu, M.S. Rehman, J.I. Han, Electricity generation and microalgae cultivation in microbial fuel cell using microalgae-enriched anode and bio-cathode, Energy Convers. Manag. 79 (2014) 674–680.

[45] S. Gupta, A. Nayak, C. Roy, A.K. Yadav, An algal assisted constructed wetland-microbial fuel cell integrated with sand filter for efficient wastewater treatment and electricity production, Chemosphere 263 (2021) 128132.

[46] R. Kakarla, B. Min, Photoautotrophic microalgae *Scenedesmus obliquus* attached on a cathode as oxygen producers for microbial fuel cell (MFC) operation, Int. J. Hydrog. Energy 39 (2014) 10275–10283.

[47] S. Kondaveeti, K.S. Choi, R. Kakarla, B. Min, Microalgae *Scenedesmus obliquus* as renewable biomass feedstock for electricity generation in microbial fuel cells (MFCs), Front. Environ. Sci. Eng. 8 (2014) 784–791.

[48] A.E. Inglesby, K. Yunus, A.C. Fisher, In situ fluorescence and electrochemical monitoring of a photosynthetic microbial fuel cell, Phys. Chem. Phys. 15 (2013) 6903–6911.

[49] L. Gouveia, C. Neves, D. Sebastião, B.P. Nobre, C.T. Matos, Effect of light on the production of bioelectricity and added-value microalgae biomass in a photosynthetic alga microbial fuel cell, Bioresour. Technol. 154 (2014) 171–177.

[50] W. Logroño, M. Pérez, G. Urquizo, A. Kadier, M. Echeverría, C. Recalde, G. Rákhely, Single chamber microbial fuel cell (SCMFC) with a cathodic microalgal biofilm: a preliminary assessment of the generation of bioelectricity and biodegradation of real dye textile wastewater, Chemosphere 176 (2017) 378–388.

[51] S.B. Velasquez-Orta, T.P. Curtis, B.E. Logan, Energy from algae using microbial fuel cells, Biotechnol. Bioeng. 103 (6) (2009) 1068–1076.

[52] A. Khandelwal, A. Vijay, A. Dixit, M. Chhabra, Microbial fuel cell powered by lipid extracted algae: a promising system for algal lipids and power generation, Bioresour. Technol. 247 (2018) 520–527.

[53] M. Shukla, S. Kumar, Algal growth in photosynthetic algal microbial fuel cell and its subsequent utilization for biofuels, Renew. Sust. Energ. Rev. 82 (2018) 402–414.

[54] S. Arun, A. Sinharoy, K. Pakshirajan, P.N. Lens, Algae based microbial fuel cells for wastewater treatment and recovery of value-added products, Renew. Sust. Energ. Rev. 132 (2020) 110041.

[55] Y. Asensio, I.B. Montes, C.M. Fernandez-Marchante, J. Lobato, P. Cañizares, M.A. Rodrigo, Selection of cheap electrodes for two-compartment microbial fuel cells, J. Electroanal. Chem. 785 (Suppl. C) (2017) 235–240.

[56] J.Y. Wu, C.H. Lay, S.R. Chia, K.W. Chew, P.L. Show, P.H. Hsieh, C.C. Chen, Economic potential of bioremediation using immobilized microalgae-based microbial fuel cells, Clean Techn. Environ. Policy 23 (8) (2021) 2251–2264.

[57] N. Rashid, Y.F. Cui, S.U.R. Muhammad, J.I. Han, Enhanced electricity generation by using algae biomass and activated sludge in microbial fuel cell, Sci. Total Environ. 456–457 (2013) 91–94.

[58] R.S. Berk, J.H. Canfield, Bio-electrochemical energy conversion, Appl. Microbiol. 12 (1964) 10–12.

[59] A. Zabihallahpoor, M. Rahimnejad, F. Talebnia, Sediment microbial fuel cells as a new source of renewable and sustainable energy: present status and future prospects, RSC Adv. 5 (114) (2015) 94171–94183.

[60] A.S. Commault, G. Lear, P. Novis, R.J. Weld, Photosynthetic biocathode enhances the power output of a sediment-type microbial fuel cell, N. Z. J. Bot. 52 (1) (2014) 48–59.

[61] X. Song, W. Wang, X. Cao, Y. Wang, L. Zou, X. Ge, Y. Wang, Chlorella vulgaris on the cathode promoted the performance of sediment microbial fuel cells for electrogenesis and pollutant removal, Sci. Total Environ. 728 (2020) 138011.

[62] X. Wang, Y. Feng, J. Liu, H. Lee, C. Li, N. Li, N. Ren, Sequestration of CO2 discharged from anode by algal cathode in microbial carbon capture cells (MCCs), Biosens. Bioelectron. 25 (12) (2010) 2639–2643.

[63] A.H. Alami, S. Alasad, M. Ali, M. Alshamsi, Investigating algae for CO2 capture and accumulation and simultaneous production of biomass for biodiesel production, Sci. Total Environ. 759 (2021) 143529.

[64] X. Wu, X. Xiong, G. Owens, G. Brunetti, J. Zhou, X. Yong, X. Xie, L. Zhang, P. Wei, H. Jia, Anode modification by biogenic gold nanoparticles for the improved performance of microbial fuel cells and microbial community shift, Bioresour. Technol. 270 (2018) 11–19.

[65] M.E. Elshobary, H.M. Zabed, J. Yun, G. Zhang, X. Qi, Recent insights into microalgae-assisted microbial fuel cells for generating sustainable bioelectricity, Int. J. Hydrog. Energy 46 (4) (2021) 3135–3159.

[66] K.S. Aiyer, Synergistic effects in a microbial fuel cell between co-cultures and a photosynthetic alga Chlorella vulgaris improve performance, Heliyon 7 (1) (2021) e05935.

[67] L. Xiao, E.B. Young, J.A. Berges, Z. He, Integrated photo-bio-electrochemical system for contaminants removal and bioenergy production, Environ. Sci. Technol. 46 (698) (2012) 11459–11466.

[68] Y. Sun, G. Yu, G. Xiao, Z. Duan, C. Dai, J. Hu, X. Jiang, Enhancing CO2 photo-biochemical conversion in a newly-designed attached photo-bioreactor characterized by stacked horizontal planar waveguide modules, Sci. Total Environ. 760 (2021) 144041.

[69] S. Ge, M. Madill, P. Champagne, Use of freshwater macroalgae Spirogyra sp. for the treatment of municipal wastewaters and biomass production for biofuel applications, Biomass Bioenergy 111 (2018) 213–223.

[70] T.K. Marella, A. Datta, M.D. Patil, S. Dixit, A. Tiwari, Biodiesel production through algal cultivation in urban wastewater using algal floway, Bioresour. Technol. 280 (2019) 222–228.

[71] N. Neveux, M. Magnusson, L. Mata, A. Whelan, R. de Nys, N.A. Paul, The treatment of municipal wastewater by the macroalga *Oedogonium sp.* and its potential for the production of biocrude, Algal Res. 13 (2016) 284–292.

[72] N. Kannan, P. Donnellan, Algae-assisted microbial fuel cells: a practical overview, Bioresour. Technol. Rep. 100747 (2021).

[73] M.J. Vucko, A.J. Cole, J.A. Moorhead, J. Pit, R. de Nys, The freshwater macroalga Oedogonium intermedium can meet the nutritional requirements of the herbivorous fish *Ancistrus cirrhosus*, Algal Res. 27 (2017) 21–31.

[74] J. Li, J. Hu, C. Yang, W. Pu, H. Hou, J. Xu, B. Liu, J. Yang, Enhanced detection of toxicity in wastewater using a 2D smooth anode based microbial fuel cell toxicity sensor, RSC Adv. 9 (2019) 8700–8706.

[75] Y. Duan, F. Shi, Bioreactor design for algal growth as a sustainable energy source, in: Reactor and Process Design in Sustainable Energy Technology, Elsevier B.V, 2014, pp. 27–60.

[76] R.G. Saratale, C. Kuppam, A. Mudhoo, G.D. Saratale, S. Periyasamy, G. Zhen, G. Kumar, Bioelectrochemical systems using microalgae–a concise research update, Chemosphere 177 (2017) 35–43.

CHAPTER 5

Prospects of algal bioactive compounds in the cosmetic industry

Anjali Yadav[a], Sumit Kumar[a], Bansal Deepak[b], Rachna Bhateria[c], and Sharma Mona[d]
[a]Department of Environmental Science and Engineering, Guru Jambheshwar University of Science & Technology, Hisar, Haryana, India
[b]JBM Group, Gurugram, Haryana, India
[c]Department of Environmental Sciences, Maharshi Dayanand University, Rohtak, Haryana, India
[d]Department of Environmental Studies, Central University of Haryana, Mahendergarh, Haryana, India

1 Introduction

Algae are members of the kingdom Protista [1] and are simple, nonflowering, and typically aquatic plants of a huge group of photosynthetic organisms that extend from the microscopic and unicellular (microalgae) to large and multicellular seaweeds (macroalgae) [2]. Macroalgae are further categorized into three main groups because of their color-pigmentation: Chlorophyceae, Rhodophyceae, and Phaeophyceae. Algae are largely used because of the diverse variety of bioactive compounds, minerals, and vitamins in their composition [3]. Many benefits are also associated with the wide variety of bioactive compounds [4]. Its benefits are clear from food to medicine to skin care to biofuels as well as wastewater treatment [5].

Algal species are the undebatable riches of the ocean [6]. Most harsh and hard living surroundings and environments do not create any troublesome problems for the life and growth of algae because of their huge adaptability [7]. Algae from every corner of the ocean are on the rise in cosmetics and personal care product formulations [8]. Over the last several years, various kinds of research have been carried out about the composition of nutritional components, pharmaceutical applications, secondary metabolites and bioactive compounds, potential health-promoting effects, and antioxidant and antibacterial properties among many others [3,5,8–10]. Due to this, the cosmetic industry is keen for utilizing algae as a source of biologically active raw material because they are highly rich in bioactive compounds as well as natural, cheaper, and healthier which are auxiliary advantages for the industry [11,12]. Algae species can produce biologically active compounds with budding future applications in the cosmetic industry and these types of applications are associated with thickening, water-binding, and antioxidant agents [13].

Algae Based Bioelectrochemical Systems for Carbon Sequestration, Carbon Storage, Bioremediation and Bioproduct Generation
https://doi.org/10.1016/B978-0-323-91023-1.00006-5

2 Bioactive compounds used for the production of cosmetics

Algae produce a variety of biologically active compounds, which have great future applications. The bioactive compounds available in algae (Table 1) are primary metabolites (polysaccharides, proteins, fatty acids, and amino acids) and secondary metabolites (vitamins, pigments, sterols, phenolic compounds, and other bioactive agents) and these algal compounds have direct aptness in cosmetics [14]. Algae provides the cosmetic industry with bioactive compounds like fucoidans, fucans, ulvans, carrageenans, tocopherol, vitamins, pigments, flavonoids, astaxanthin, etc., which have shown immense benefits to human skin [15]. The most prominent reason for the use of algae in cosmetics is because it is exceptionally rich in nutrients and minerals like K, Mg, Ca, and vitamins A, vitamins B-12 [16]. Therefore, when this natural power is harnessed and popped into skin care, hair care, and beauty products, it can deliver dynamic results [17]. The biologically active substances present in algae can be identified and extracted, and these compounds can function as nutritional supplements, cosmeceuticals, cosmetics, and enzymes for various products [18].

3 Skin benefits of algal metabolites

In the human body, the skin is the most complex and largest organ that constitutes about one-sixth of the total body weight; it has various functions to perform. The most essential one is to protect the body against harmful chemicals, pathogens, UV radiation, and various other physical stressors. With increasing age and a harmful and stressful environment, the skin becomes more vulnerable to natural processes like skin aging, pigmentation, wrinkling, dullness, and loss of suppleness and smoothness. Here, the algal products appear as they fight against the harmful external components that threaten the skin, and it is natural, healthier, and affordable as compared to other cosmetic products as shown in Fig. 1 [28,29]. The cosmetic industry could merely survive in the absence of algae as the nutrient density, that is, amino acids, mineral salts, trace elements, and vitamins are quite like the human body and its system [30]. It promotes blood circulation, supplies moisturizer to the skin, and synchronizes the function of the sebaceous gland. They initiate cell resumption and metabolism, escalate the skin's resilience, and have an antiinflammatory outcome. The proteins present in algae supply the skin epidermis with energy, mucus substances safeguard the skin from drying off, and vitamins trigger the skin and secure it against natural changes [29,31]. Chlorophyll remarkably puts up to the oxygen demand of the skin as it is alike to the hemoglobin. These ample qualities are utilized in skin care formulas for dry and rough skin or with diminished flexibility [32].

Table 1 Types of bioactive compounds obtained from given algal species.

Bioactive compound	Types	Algae species	Reference
Polysaccharides	Alginate, fucoidan, laminarin, carrageenan, cellulose, furcellaran, mannan, porphyran, xylon amylose, amylopectin, insulin, mannan, pectin, xylan, ulvan	*Porphyridium* sps., *Costaria costata, Ulva lactuca, Isochrysis galbana*	[19]
Lipids and fatty acid	PUFAs, omega-3 and omega-6 fatty acids, (linolenic acid, arachidonic acid), palmitic acid	*Tetraselmis* sps., *Nannochloropsis* sps., *Porphyridium* sps.	[20]
Pigments	Phycobilins (phycocyanin, phycoerythrin) Carotenoids: carotene (alpha-carotene, beta-carotene, lycopene), xanthophyll (astaxanthin, fucoxanthin, zeaxanthin, lutein) Chlorophylls: (chlorophyll *a*, chlorophyll *b*, chlorophyll *c*)	*Gracilaria gracilis, Spirulina platensis, Dunaliella salina, Sargassum siliquastrum, Odontella aurita*	[20–22]
Antioxidants	Ascorbate, glutathione, tocopherol, polyphenols, phlorotannins, catechins, flavonoids, mycosporine-like amino acids	*Fucus vesiculosus, Cystoseira nodicaulis, Ecklonia cava, Cystoseira tamariscifolia*	[23]
Proteins and amino acid	Phycobiliproteins, phycocyanobilins, phycoerythrobilins	*Porphyra* sps., *Chlamaydomonas hedleyi, Padina crassa,*	[24–26]
Vitamins	B_{12}, K, C, E, A, D	*Dunaliella tertiolecta, Chlorella* sps., *Porphyra umbilicalis, Gracilaria changii*	[10]
Minerals	K, Ca, Mg, Zn, Na, Co, B, Cu, I	*Gracilaria* sps., *Laminaria* sps., *Chondrus crispus*	[27]

4 Current trends

The cosmetic industry is flourishing day by day as awareness among people is increasing to protect their skin and keep it healthy and glowing. Today's consumer is more inclined towards natural products as the synthetic ones can cause adverse side effects. Hence, algae are an attractive source for skin care and the benefits provided by it are abundant not only in the cosmetic industry but also in the food industry, pharmaceuticals, etc. [33].

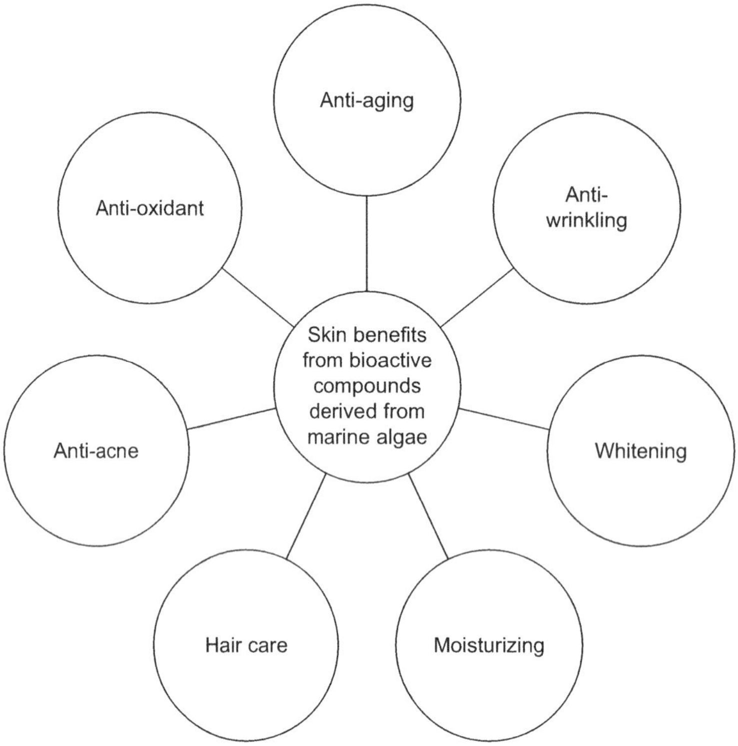

Fig. 1 Skin benefits from bioactive compounds derived from marine algae.

"There is a growing global trend towards the use of cosmetic products that are perceived as healthy and environmentally sustainable. This trend has prompted cosmetic firms to invest in research and development efforts focused on the creation of new cosmetic products that incorporate bioactive ingredients or extracts derived from natural sources." The advancement in the cosmetic industry has been compelled by the current intrigue exhibited by conscious and enlightened customers in the products obtained or synthesized from natural sources rather than synthetic ones [34]. The use of plant-derived extracts in cosmetics is showing new alternatives [35,36]. It has been noticed that the advancement of technology in the marine industry like deriving new pharmaceuticals, enhanced aquaculture, etc., causes increased use of products extracted from marine organisms, especially microalgae and seaweeds [37]. The increasing need to procure safe and secure products by environment-friendly biological processes has resulted in making algae an affordable, healthier, and sustainable resource for new natural products in the cosmetic industry [38,39]. According to a few studies, now algae become an essential source of biologically active compounds and extracts used in cosmetic products [40]. Only a few bioactive compounds are used generally in cosmetic products as many of them are not yet discovered and studied regarding their use in cosmetic formulations [41].

The global market of algal bioactive compounds is estimated to progress at a CAGR of 4.0% from 2019 to 2025 to reach USD 452.4 million by 2025. In India, this business is expected to progress at a CAGR of 4.23% in the course of the estimated span of 2020–25 [42]. The rising inclination to use natural and green products has added to the sharp demand for eco-friendly cosmetic products. The cosmetic industry is concerned with finding novel biological and natural alternatives, which can increase the quality of the products and can replace chemical compounds [43].

5 Applications of algal metabolites in cosmetics

From algae, several secondary metabolites are derived which are well known for their skin benefits [44]. Cosmetic products do not alter the physiological state of the skin but they are proposed to enhance skin appearance, boost attractiveness, and modify aspects. The bioactive compounds when merged with cosmetic formulations provide various benefits to the skin such as antiaging, depigmentation, and skin whitening, moisturize the skin, and make it smooth, supple, and glowing. Cosmetics is a domain in which algae (macroalgae and microalgae) have been strengthening their existence, and it is feasible to procure a broad variety of products like shampoos, sunscreens, antiaging creams, moisturizers, etc., for over-the-skin application.

Antiaging: Aging is a natural process of getting older; it means that the organism's system becomes slow in repairs, responses, and regeneration. It is an intricate biological activity that generally refers to losing skin elasticity, the emergence of fine lines, and discoloration or pigmentation of the skin with increasing age. The most prominent result of aging is seen on the skin in humans. Even though it is a natural process but it is escalated due to extremely harmful environmental factors, exposure to UV radiations, heavy metals, lack of nutrients, and moisture on the epidermis. Thus, arise problems like skin laxity, thinning, enlarged pores, fragility, losing fat, reduced skin elasticity, plumpness and smoothness of skin, and discoloration of the skin. The psycho-social influence of skin aging resulted in great demand for cosmetics with antiaging properties. To overcome these problems of skin aging, various antiaging skin care products and treatments are made available. Algae are used in cosmetic products owing to the presence of vitamin C, E, β-carotene, antioxidants, and other beneficial compounds in them, which help against skin aging and make it immune to signs of aging. Algal bio-compounds are efficient in increasing the activities to restrain the action of hyaluronidase, matrix metalloproteinase (MMP), and elastase inhibitory activities, and collagen expression, might have the capacity to be utilized as biologically active ingredients in new and unique cosmetic products with attributes that reduce aging signs [30]. Applying fucoidan over the skin has even been revealed to possess antiaging action by enhancing elasticity and moistening cells. Algae used in the production of antiaging products: *Porphyra umbilicalis, Chlorella,*

Dunaliella, Arthospira, Porphyridium, Chalmydocapsa, Gracilaria, Thalassiosira, Polysiphonia, Colpomenia, Halymenia, Padina, Ahnfeltiopsis, etc.

Skin whitening: Also known as skin lightening or skin bleaching, it is an attempt to lighten the skin color, remove tanning (constant exposure to sunlight), or obtain even skin color by reducing the concentration of melanin in the skin. Skin-whitening cosmetics appear to happen the greatest and continuously flourishing section in the cosmetics market. Generally, chemical compounds are used for the same, but natural and healthier algal products are replacing these toxic substances. The presence and concentration of melanin produced in humans determine the color of skin and it also acts as a protective barrier for the skin from harmful UV radiation, and scavenging toxic substances. Continuous and long-time exposure to sunlight induces the synthesis of tyrosinase, which further catalyzes the reaction for melanin production. The considerable amount of melanin results in skin pigmentation, which requires to be regulated. Algal products give the result by blocking the functioning of the tyrosinase enzyme responsible for melanin production in the skin. Fucoxanthin and zeaxanthin are known to help reduce tyrosinase activity [16]. Algae used in skin-whitening products: *Nannochloropsis oculata, Alaria, Chorda, Macrocystis, Laminaria japonica, Astaxanthin, Halidrys siliquosa, Ascophyllum nodosum, Palmaria palmata,* etc.

Moisturizing agent: Moisturizer is a cosmetic preparation used for skin protection, moisturizing, and lubrication. Moisturizing products help to enhance and support the barrier function of the skin, retaining it softer with a thriving appearance, and preventing it from dryness, acne, bruising, and wrinkling. If the skin is not moisturized properly, it is prone to dryness, aggravated acne, and wrinkling, and can even result in eczema [16]. Algal bio-compounds like fucoidans, agar, alginate, and carrageenan are nonirritating, nontoxic, have emollient properties, and as well are abundant and economical for use as an alternative in cosmetic products. Some hydrolyzates or algal proteins show sturdy chemistry with providing moisturizer and water retention to the skin soothing effect. Algal species used in the processing of moisturizers: *Thraustochytrium, Chondrus crispus, Codium tomentosum, Aurantiochytrium, Porphyra, Spirulina, Chlorella, Fucus vesiculosus, Nostoc commune.*

Hair care: Hair care is a collective term used for hygiene and cosmetology involving the hair on the human scalp. Hair care routine varies from person to person and even in culture and regions; it includes cleaning, oiling, coloring hair, preventing hair loss, etc. Hair loss is an increasing problem in today's world and this can be due to various reasons like pollution, UV radiation, chemical substances, pathological enhancement, aging, etc.; to prevent hair loss, hair fall, and hair damage is not easy [11]. Algae used in the production of hair care products: *Porphyridium cruentum, Chlorella, Monodus, Thalassiosira, Chaeloceros.*

6 Conclusion

Algae, as well as, algal products, can be regarded as potential and valuable resources for human cosmeceuticals. Algae are now diligently used as a substitute in various cosmetic formulations as one of the most reliable organic ingredients. They produce many bioactive compounds like carotenoids, polysaccharides, proteins, pigments, vitamins, and minerals during their growth. Because of this specific attribute, they are regarded as an essential natural resource for cosmetic formulations. These metabolites have diverse functions and applications. They are potent to fight against natural processes and external stimuli. Algal-derived metabolites provide various skin benefits like antiaging, whitening, moisturizing, and antiwrinkling. Therefore, adapting to products derived from algae is not only sustainable but also good for human use. The future perspectives can include the active characterization and their overall action of bioactive compounds jointly and separately and cytotoxic studies also need to be done.

References

[1] R.A. Aendersen, R.A. Lewin, Algae, 2019, (Date accessed August 14, 2020) Retrieved from https://www.britannica.com/science/algae.

[2] www.lexico.com/definition/alga, 2020..

[3] R. Sathasivam, R. Radhakrishnan, A. Hashem, E.F. Abd-Allah, Microalgae metabolites: a rich source for food and medicine, Saudi J. Biol. Sci. 26 (4) (2019) 709–722.

[4] K. Thiyagarasaiyar, B.H. Goh, Y.J. Jeon, Y.Y. Yow, Algae metabolites in cosmeceutical: an overview of current applications and challenges, Mar. Drugs 18 (6) (2020) 323.

[5] I. Ragusa, G.N. Nardone, S. Zanatta, W. Bertin, E. Amadio, Spirulina for skin care: a bright blue future, Cosmetics 8 (1) (2021) 7.

[6] N. Thajuddin, D. Dhanasekaran, Algae: Organisms for Imminent Biotechnology, 2016, ISBN: 978-953-51-2432-0.

[7] S.M. Gärtner, B. Uzunov, G. Gärtner, Enigmatic microalgae from aeroterrestrial and extreme habitats in cosmetics: the potential of the untapped natural sources, Cosmetics 7 (2) (2020) 27.

[8] T. Morais, J. Cotas, D. Pacheco, L. Pereira, Seaweeds compounds: an ecosustainable source of cosmetic ingredients, Cosmetics 8 (1) (2021) 8.

[9] S.L. Holdt, S. Kraan, Bioactive compounds in seaweed: functional food applications and legislation, J. Appl. Phycol. 23 (3) (2011) 543–597.

[10] S.S. Kumar, V. Kumar, R. Kumar, S.K. Malyan, N.R. Bishnoi, Ferrous sulfate as an in-situ anodic coagulant for enhanced bioelectricity generation and COD removal from landfill leachate, Energy 176 (2019) 570–581.

[11] L. Pereira, Seaweeds as source of bioactive substances and skin care therapy—cosmeceuticals, algotherapy, and thalassotherapy, Cosmetics 5 (4) (2018) 68.

[12] E.A. Siahaan, R. Pangestuti, H. Munandar, S.K. Kim, Cosmeceutical's properties of sea cucumbers: prospects and trends, Cosmetics 4 (3) (2017) 26.

[13] T. Wang, R. Jónsdóttir, H. Liu, L. Gu, H.G. Kristinsson, S. Raghavan, G. Ólafsdóttir, Antioxidant capacities of phlorotannins extracted from the brown algae *Fucus vesiculosus*, J. Agric. Food Chem. 60 (23) (2012) 5874–5883.

[14] L. Novoveská, M.E. Ross, M.S. Stanley, R. Pradelles, V. Wasiolek, J.F. Sassi, Microalgal carotenoids: a review of production, current markets, regulations, and future direction, Mar. Drugs 17 (11) (2019) 640.

[15] P. Kiuru, M.V. D′ Auria, C.D. Muller, P. Tammela, H. Vuorela, J. Yli-Kauhaluoma, Exploring marine resources for bioactive compounds, Planta Med. 80 (14) (2014) 1234–1246.

[16] S. Joshi, R. Kumari, V.N. Upasani, Applications of algae in cosmetics: an overview, Int. J. Innov. Res. Sci. Eng. Technol. 7 (2) (2018) 1269.

[17] T.F. Alves, M. Morsink, F. Batain, M.V. Chaud, T. Almeida, D.A. Fernandes, P. Severino, Applications of natural, semi-synthetic, and synthetic polymers in cosmetic formulations, Cosmetics 7 (4) (2020) 75.

[18] S.A. Pomponi, The bioprocess-technological potential of the sea, Prog. Ind. Microbiol. 35 (1999) 5–13.
[19] M.F. Raposo, R.M. de Morais, A.M. Bernardo de Morais, Bioactivity and applications of sulphated polysaccharides from marine microalgae, Mar. Drugs 11 (2013) 233–252.
[20] M. Francavilla, M. Franchi, M. Monteleone, C. Caroppo, The red seaweed *Gracilaria gracilis* as a multi products source, Mar. Drugs 11 (2013) 3754–3776.
[21] M.K.N. Chidambara, A. Vanitha, J. Rajesha, S.M. Mahadeva, P.R. Sowmya, G.A. Ravishankar, In vivo antioxidant activity of carotenoids from *Dunaliella salina*—a green microlalga, Life Sci. 76 (2005) 1381–1390.
[22] J. Liu, Z. Sun, H. Gerken, Z. Liu, Y. Jiang, F. Chen, *Chlorella zofingiensis* as an alternative microalgal producer of astaxanthin: biology and industrial potential, Mar Drugs 12 (2014) 3487–3515.
[23] E. Talero, M.S. Garcia, R.J. Avila, L.A. Rodriguez, A. Alcaide, V. Motilva, Bioactive compounds isolated from microalgae in chronic inflammation and cancer, Mar. Drugs 13 (2015) 6152–6209.
[24] Q. Gao, F. Garcia-Pichel, Microbial ultraviolet sunscreens, Nat. Rev. Microbiol. 9 (2011) 791–802.
[25] A. Singh, M. Čížková, K. Bišová, M. Vítová, Exploring mycosporine-like amino acids (MAAs) as safe and natural protective agents against UV-induced skin damage, Antioxidants 10 (5) (2021) 683.
[26] S.S. Suh, J. Hwang, M. Park, H.H. Seo, H.S. Kim, J.H. Lee, Anti-inflammation activities of MAAs in response to UV radiation suggest potential anti-skin aging activity, Mar. Drugs 12 (2014) 5174–5187.
[27] A.R. Circuncisao, M.D. Catarino, S.M. Cardoso, A.M.S. Silva, Minerals from macroalgae origin: health benefits and risks for consumers, Mar. Drungs 16 (2018) 400.
[28] A. Aslam, A. Bahadar, R. Liaqat, M. Saleem, A. Waqas, M. Zwawi, Algae as an attractive source for cosmetics to counter environmental stress, Sci. Total Environ. 144905 (2021).
[29] J.Y. Berthon, R. Nachat-Kappes, M. Bey, J.P. Cadoret, I. Renimel, E. Filaire, Marine algae as attractive source to skin care, Free Radic. Res. 51 (6) (2017) 555–567.
[30] M.B. Ariede, T.M. Candido, A.L.M. Jacome, M.V.R. Velasco, J.C.M. de Carvalho, A.R. Baby, Cosmetic attributes of algae—a review, Algal Res. 25 (2017) 483–487.
[31] H. Chakdar, S. Pabbi, Algal pigments for human health and cosmeceuticals, Algal Green Chemistry (2017) 171–188.
[32] H.M.D. Wang, C.C. Chen, P. Huynh, J.S. Chang, Exploring the potential of using algae in cosmetics, Bioresour. Technol. 184 (2015) 355–362.
[33] J. Rumin, E. Nicolau, R. Gonçalves de Oliveira Jr., C. Fuentes-Grünewald, L. Picot, Analysis of scientific research driving microalgae market opportunities in Europe, Mar. Drugs 18 (5) (2020) 264.
[34] D.I. Resende, M. Ferreira, C. Magalhães, J.S. Lobo, E. Sousa, I.F. Almeida, Trends in the use of marine ingredients in anti-aging cosmetics, Algal Res. 55 (2021) 102273.
[35] A.S. Ribeiro, M. Estanqueiro, M.B. Oliveira, J.M. Sousa Lobo, Main benefits and applicability of plant extracts in skin care products, Cosmetics 2 (2) (2015) 48–65.
[36] S. Trehan, B. Michniak-Kohn, K. Beri, Plant stem cells in cosmetics: current trends and future directions, Future Sci. OA 3 (4) (2017) FSO226.
[37] A. Rotter, A. Bacu, M. Barbier, F. Bertoni, A.M. Bones, M.L. Cancela, M.I. Vasquez, A new network for the advancement of marine biotechnology in Europe and beyond, Front. Mar. Sci. 7 (2020) 278.
[38] I. Michalak, K. Chojnacka, Algal extracts: technology and advances, Eng. Life Sci. 14 (2014) 581–591.
[39] E.J. Yun, I.G. Choi, K.H. Kim, Red macroalgae as a sustainable resource for bio-based products, Trends Biotechnol. 33 (5) (2015) 247–249.
[40] M. Sanghvi, Y. Martin Lo, Present and potential industrial applications of macro-and microalgae, Recent Pat. Food, Nutr. Agric. 2 (3) (2010) 187–194.
[41] V. Vaibhav, S. Sahasrabuddhe, Blue is the new green for cosmetic industry, 3 (2018) 10.
[42] www.meticulousresearch.com.
[43] S. Santhosh, R. Dhandapani, N. Hemalatha, Bioactive compounds from microalgae and its different applications-a review, Adv. Appl. Sci. Res. 7 (4) (2016) 153–158.
[44] R. Pangestuti, S.K. Kim, Biological activities and health benefits of natural pigments derived from marine algae, J. Funct. Foods 3 (2011) 255–266.

CHAPTER 6

Algae-based bio-electrochemical systems for carbon sequestration, bioremediation, and cogeneration of valuable chemicals: Challenges and future prospects

Jagdeep Kumar Nayak[a,b], Rahul Gautam[a,c], Chung-Yu Guan[d], and Uttam Kumar Ghosh[a]
[a]Department of Polymer and Process Engineering, Indian Institute of Technology, Roorkee, India
[b]Department of Civil and Architectural Engineering, Sultan Qaboos University, Muscat, Oman
[c]Department of Chemical Engineering, University of Birmingham, Birmingham, United Kingdom
[d]Department of Environmental Engineering, National Ilan University, Yilan, Taiwan

1 Introduction

Economic development, social growth, and quality of life require energy as the main element to match with the rapid urbanization, industrialization, and modernization, which are possible due to energy supply. The major energy demands meets by finite fossil reserves. These soaring fossil fuel prices and high energy demands along with the associated environmental impacts, e.g., increasing CO_2 concentration, global warming, and climate change, etc. have challenged the scientific panel to seek for an alternative source of energy to take on the increasing costs of fossils fuels and high energy demands. It is the current hotspot of the research [1–7]. Renewable energy with fewer emissions is also cost-efficient. It ensures energy security by decreasing dependence on fossil fuels [5,8–11]. Energy derived from biomass, fuel cell technology, geothermal energy, tidal energy, solar energy, etc. are types of renewable energy that are proposed and worked on to meet the energy shortage and greenhouse gases issues. The new approach of waste to energy also starts to come into existence as it could be the source of various energy sources [10–15]. So the need is to focus on a sustainable approach for waste treatment with simultaneous energy production to address the issue of energy scarcity and high volume of waste [16–21].

Bio-electrochemical systems, solar energy-efficient algal-based biofuels, etc. have attracted enormous attention. Both systems can be fed on the waste to extract energy from it. These systems are excellent in CO_2 sequestration too. Algae, one of the most

Algae Based Bioelectrochemical Systems for Carbon Sequestration, Carbon Storage, Bioremediation and Bioproduct Generation
https://doi.org/10.1016/B978-0-323-91023-1.00011-9

primitive organisms on earth, can be cultivated almost anywhere with minimal nutrient requirements without competing with food crops [3,22]. Microbial fuel cell (MFC) is presented as a technique for bio-electricity production by utilizing organic matter of the feed and thus results in the removal of chemical oxygen demand (COD), heavy metals, etc. The sustainable bio-electrochemical system (BES) can transform the chemical energy stored as organic matter into different usable energy forms with the help of microorganisms to oxidize the feed [23,24]. One of the major among BES is MFC technology that simultaneously treats wastewater with electricity generation by expending fewer quanta of energy when degrading contaminants under mild operation conditions [12,13]. It is rated as the transducer of the BES system where biocatalysis of the feed results in the production of electricity [25,26]. The conceptualization of the PMFC is based on using the photosynthetic microorganism in the cathode chamber of the MFC, which will sequestrate the atmospheric CO_2, or from the flue gases or from the anodic chamber and results in biomass productivity. It is meant to increase the power out and productivity of the system [26–28]. It is based on the fact that microalgal carbon sequestration has been persuaded by many countries to explore sustainability [3]; they are mitigating the carbon dioxide from various industrial sectors off-gases to reduce greenhouse gas emissions through the mass production of microalgae. Microalgal culturing is advantageous as it is rich in triglycerides and fatty acids, so acts as a feedstock for biodiesel production through the transesterification process [29]. Microalgae culture reactors are generally of two types, i.e., open and closed bioreactors. The open reactors are simple in design and construction, low in cost, and easy in operations, while these systems also offer shortcomings in operations, e.g., large area requirement, uncontrolled culture conditions, pollution susceptibility, culture medium losses due to evaporation, etc. It leads to the development of the closed type of photobioreactors, which has better control of culture conditions and offers lesser evaporation losses. The controllable culture conditions in the closed-type culture systems make them efficient systems. They are also known as photobioreactors, and have high production, fewer contaminations, and less water consumption, but they incur high construction and operational costs [3]. The microalgal biomass-based carbon dioxide-sequestrated biofuels are preferable in comparison to the other cellulosic biomass-based biofuels [30,31]. They grow in a wide variety of environmental conditions to perform CO_2 sequestration. The structural components of algae store lipids as energy reserves making it a suitable resource for biodiesel production [28]. The microalgal biomass production is the function of the temperature, illumination, light intensity, etc. [32].

1.1 Bioremediation

The unaware and unrestrained release of wastes and wastewater into the surroundings is degrading the environment as a whole. The persistent pollutants are troublesome

[33–35]. A branch of biotechnology is drawing attention, i.e., bioremediation. It is the removal of the contaminants and pollutants by using microorganisms. It is a very useful tool to degrade oil spills, to remove soil contaminants, and wastewater treatment [26,28,34]. A few of the microorganisms release the exopolysaccharides with the help of which they perform bioremediation.

The integration of microalgal carbon sequestration and pollutant removal along with the additional biofuel generation is a recent and innovative domain of the research. The bio-electrochemical systems can provide such a platform for the floriation of the concept [35].

2 Bio-electrochemical system

2.1 Microbial fuel cell technology

MFC is a self-sustainable BES based on the oxidation of the substrate aided by catalysis performed by the microorganism [10,13,14,22–24,36]. The electricity generation through bacterial oxidation of organic substrates in MFC is a unique and ideal way for simultaneous bioelectricity generation and bioremediation. The most basic design of MFC is an H-type which is a dual-chamber having an anodic chamber and a cathodic chamber which are separated by a membrane called as proton exchange membrane (PEM) (Fig. 1) [37–39]. The anodic bacterial population involves in organic matter oxidation which results in the generation of electrons, protons, and CO_2. The generated electrons are first accepted by nicotinamide adenine dinucleotide (NADH) to transfer these electrons to oxygen, sulfate, nitrate, etc. as the terminal electron acceptors [37]. These electrons reach the anode electrode surface through the external circuit and finally transfer to the cathode electrode [40,41].

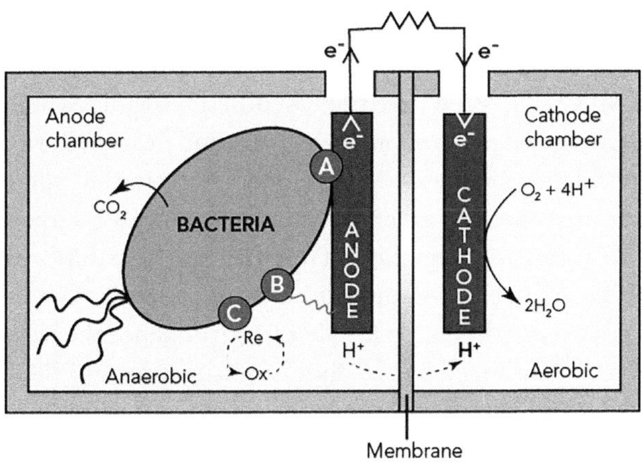

Fig. 1 Principle of MFC (Electrons can be carried by protein carriers (A), nanowires (B) or mediators (C)) [41].

Diffusion is the process, which enables the H^+ transfer from anode to cathode through PEM such as Nafion, Ultrix, and salt bridge. The protons and the electrons at the cathode combine with the oxygen resulting in the production of water. The anodic chamber is maintained anaerobic, while the cathode chamber is kept under aerobic conditions [37,42]. The Eqs. (1) and (2) are the reactions taking place at the MFC electrodes using acetate as substrate.

At anode:

$$CH_3COO^- + 2H_2O \rightarrow 2CO_2 + 7H^+ + 8e^- \quad (1)$$

At cathode:

$$O_2 + 4e^- + 4H^+ \rightarrow 2H_2O \quad (2)$$

The potential difference between the electrodes serves as the driving force for smooth electron transfer between the electrodes. Under standard operating conditions, the maximum potential difference of about ~1.2 V between NADH (the electron donor) and the best electron acceptor oxygen can be observed. Wastewater is a source of a wide range of inorganic matter and organic matter, and thus can be explored as a fuel. The oxygen has high oxidation potential, thus regarded as the best electron acceptor to produce water as a byproduct [43]. Other chemicals have also been studied as an electron acceptor, e.g., potassium ferricyanide, platinized cathode, etc., to increase the productivity but lead to higher operation cost and toxicity, respectively [44].

The contaminates like chlorides, sulfates, arsenates, nitrates, etc., in the wastewater can also act as an electron acceptor and result in simultaneous contaminant removal [37,45–48]. Taking these beneficial aspects of MFC into account, this sustainable technology can be put forward as a tool for simultaneous renewable energy production and bioremediation.

2.2 Algal-based MFC

PMFC is a type of MFC (Fig. 2), where photosynthetic organisms are recruited for oxidation, charge separation, and generation of electrons and protons during flow chain reactions. The symbiotic relationship between algae and heterotrophic microorganisms results into an efficient treatment process where oxygen is produced by microalgae using solar energy and nutrients. With this integration of microalgae, the traditional treatment system became photosynthetic,more environmentally and economically sustainable [49]. Thus, it is assumed that the growing cathodic algae in MFC can enhance the rate of electricity generation, where oxygen is produced by the photosynthesis process [50]. Thus, algae-based photosynthetic MFC represents a more sustainable advanced technique as compared to the typical MFCs for simultaneous CO_2 sequestration, biomass production, electricity generation, and wastewater treatment [51,52].

In PMFC, the algal biocathode as a mediator has been employed as an electron acceptor for the conversion of CO_2 to biomass. The algal cells consume the electron for the

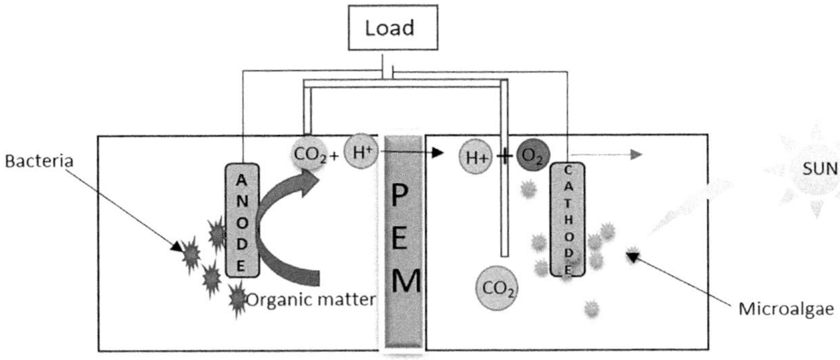

Fig. 2 Schematic diagram of PMFC.

metabolic activities for their growth and results in biomass and oxygen (Figs. 1 and 2). The algal cells release the oxygen as the oxidized mediator and the mediator again gets reduced in the catholyte by the transferred electron [53]. The following reactions are taking place in both the cathodic and anodic chambers as mentioned in Eqs. (3), (4), respectively [54].

$$6CO_2 + 12H^+ + 12e^- \rightarrow C_6H_{12}O_6 \text{ (biomass)} + 3O_2 \quad (3)$$

$$C_6H_{12}O_6 + 6H_2O \rightarrow 6CO_2 + 24H^+ + 24e^- \quad (4)$$

$$C_6H_{12}O_6 + 6O_2 \rightarrow 6CO_2 + 6H_2O \quad (5)$$

In the presence of light, algae carry out photosynthesis with the help of CO_2 to produce biomass and organic matter, while in dark reactions they consume oxygen and oxidize the formerly produced organic matter to get energy (Eq. 5) [55]. Similar design and work have been done by Wang et al. (2010), where they demonstrated the setup called microbial carbon capture cell (MCC) in which the generated CO_2 from substrate degradation was transferred from the anodic chamber to the cathodic chamber for the utilization in algal photosynthesis [56]. This process represents a sustainable technology for both CO_2 sequestration for biomass generation and bioelectricity generation. However, this technology is still in its beginning stage. A lot of work needs to be done for its full-fledged implementation. There are studies available on some photosynthetic cyanobacteria such as *Spirulina platensis* which were used as bio-anode catalysts, whereas the algae helped to maintain the electrochemical potential with the formation of biofilm which can accept the electrons directly [57,58].

3 Microalgal carbon sequestration: Mechanism and value-added products

3.1 Carbon sequestration

The anthropogenic increase in CO_2 concentration in the atmosphere is causing global warming and an average increase in the earth's temperature, which is approaching a

2°C rise. The global CO_2 concentration has reached to 420 ppm. This is an alarming situation where the technologies to capture and store of CO_2 are required. This process of capturing and storage of CO_2 is termed is as carbon sequestration; it will be mitigated by curbing on emissions and simultaneously capturing and storage [59]. Carbon sequestration can be achieved either naturally or anthropogenically or either biotic or abiotic [60,61]. This can be achieved by many processes—the restoration of the forest cover is the best and natural method, while others are carbon farming, geological and oceanic injection, production of biochar, and ocean farming where seaweeds which mainly consist of micro- and macroalgae perform the photosynthesis, etc. [33,59–61]. The algal-mediated C sequestration is a biotic mode of sequestration, and productivity is higher than the green plants.

BES coupled with microalgae for CO_2 sequestration is a promising and innovative approach with multiple outcomes such as CO_2 capture, bioenergy (bioelectricity and biodiesel) generation and wastewater treatment [10,51,62]. The anaerobic digestion in the anodic chamber by exoelectrogens produces protons, electrons, and CO_2. If this anodic CO_2 is transferred to the cathodic chamber equipped with the photosynthetic microorganism (e.g., algae, cyanobacteria), then it converts CO_2 to biomass in the presence of a source of energy, i.e., light illumination [13,63]. The research efforts are mainly focused on performance enhancement by optimizing the cathodic configuration and operating conditions for excellent end products.

3.2 CO_2 sequestration mechanism in microalgae

Microalgal carbon sequestration has been considered auspicious, as it along with photosynthesis produces other value-added products. It is a two-step process composed of a carbon-concentrating mechanism and a photosynthesis process, the most vital characteristics of autotrophs [29,64]. Likewise, plants' photosynthesis and microalgal photosynthesis take place in two steps, i.e., light reaction and dark reaction [51]. The light reaction is characterized by the absorption of photons from sunlight bychlorophyll to excite the electrons of the water molecule. It results in water splitting to release the molecular oxygen and energy in the form of NADPH and ATP. In the dark reaction, the sugar is produced by the sequestration of the carbon dioxide and utilization of the energy produced in light reactions. Most of the algae follow C3 cycle or Kelvin cycle for the fixation of the CO_2 into sugars. The Kelvin cycle involves the carbon fixation, where CO_2 binds with the five carbon molecule, Ribulose 1,5 bisphosphate (RuBP) in the presence of ribulose bisphosphate carboxylase (RuBP) to produce three carbon compound 3-phosphoglycerate (3 PGA) and then to another 3 carbon compound glyceraldehyde 3-phosphate (G3P) [28,32]. The two molecules of G3P are believed to produce one molecule of glucose, a six carbon compound that breaks down to provide energy to organisms (Eq. 6) [28]. It is estimated that a microsized algae can fix double its weight of atmospheric carbon dioxide (Fig. 3) [28,65]. To improve productivity in terms of lipid

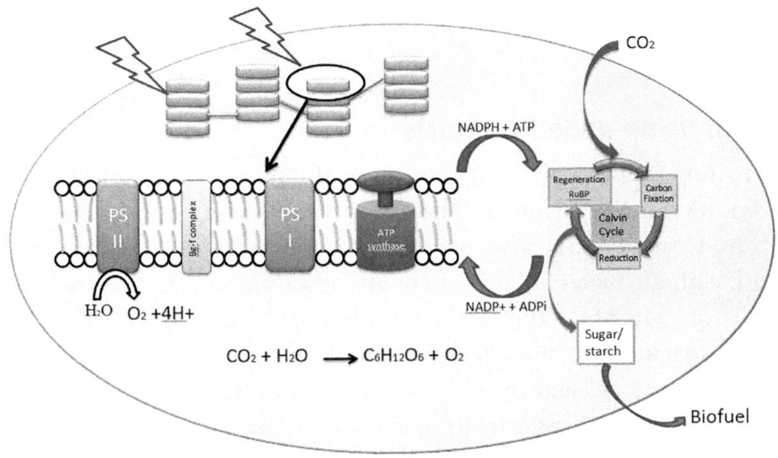

Fig. 3 Mechanism of carbon sequestration.

accumulation, algae are being sequenced and subjected to genetic modifications to improve the biomass yield [65].

$$6CO_2 + 6H_2O = C_6H_{12}O_6 + 6O_2 \tag{6}$$

Carbon sequestration in microalgae is affected by key parameters, e.g., light intensity, photoperiod and wavelength of the light, pH, temperature, etc. [3,25].

3.3 PMFC for CO_2 sequestration

MFCs are the class of BES; they are renewable and generate bio-electricity out of waste [26]. The cathode type and its material play an important role in determining the efficiency of the MFCs. The noble metals like platinum are costly and have environmental issues too. To limit these concerns, the biological cathodes or biocathodes are the focus of research. Algal-based biocathodes are potent to serve the aim of electricity generation along with biomass generation for different uses [66]. The most important feature of algal biocathodes is that it can aid in carbon sequestration. The atmospheric carbon dioxide, or from flue gases, or from the anodic chambers of the MFCs can be passed to the cathodic chamber, where autotrophic microalgae can entrap it to convert into biomass by the most important biological feature of photosynthesis. The photo hours and intensity of the sunlight are the crucial features for the productivity of the MFCs. The high lipid content in the autotrophic algal biomass can be explored for biodiesel, biochar, or as the feed for the bacteria in the anodic chamber. The photosynthesis ability of the microalgae is better than that of the terrestrial plants. The oxygen coming out of the cathodic chamber will be an added advantage of the process [51]. In recent years, a number of ideas are flowing about the PMFC design and configuration, but the dual-chambered design is the most common and widely studied design [26]. Microalgal-assisted cathode, a type of biocathodes, has higher efficacy and inferior costs. These algal biocathodes are excellent in the removal of contaminates too [67].

3.4 PMFC for value-added products

Algal biomass for several applications is one of the most striking products; it can be attained from the cathode chamber. The easiest way to scrap out the attached biofilm from the cathode surface [51]. It depends on many factors, e.g., cell–cell interaction, cell adhesion, cathode material, etc. There are other physical factors too which affect biofilm formation. The pH of the anodic chamber, CO_2 concentration and light intensity inside the reactor, and mass transfer conditions are few to mention. Effective biofilm of *Scenedesmus obliquus* over a platinized cathode is also being achieved [26]. Alternatively, the algal biomass can be separated by using common low-cost harvesting methods, e.g., membrane separation, flocculation, cell immobilization, etc. Recovery of value-added chemicals from microalgae in the PMFCs requires pretreatment. Table 1 summarizes the bioremediation and biomass productivity of the different algal species in different PMFCs and other systems.

A new approach of biorefinery has dual advantages out of MFC, i.e., simultaneous production of bio-electricity and value-added chemicals [25,67]. The produced biomass in PMFC has different utilities, e.g., as a source of biogas (CH_4), CO_2 as byproduct as a source of inorganic carbon, or biomass as animal feed (Fig. 4).

4 Influencing factors

4.1 Illumination and photoperiod

Among all the environmental factors, illumination and photoperiod are one of the most important factors affecting algal productivity [72]. The illumination is the driving force of

Table 1 Bioremediation and biomass productivity of different algal species.

Substrate	Algal species	Reduction	Biomass (g/L)	Reference
Pharmaceutical WW	*T. indica*	90% COD removal	0.78 ± 0.05	[12]
Distillery spent wash	S. abundans	66%–78% COD removal	0.74 g/L	[13]
Kinnow peel extract	*T. indica*	44%–95% nutrients removal	1.02 ± 0.03	[68]
Pharmaceutical WW	*S. abundans*	77%–97% COD removal	0.94 ± 0.01 g/L	[69]
Swine WW	*C. vulgaris*	96% COD removal	164 mg/L/d lipid productivity	[70]
Municipal WW	*Chlorella sorokiniana*	$NH4^+$-N 29.7%–60.6%	25.0 ± 0.1 mg/L	[71]

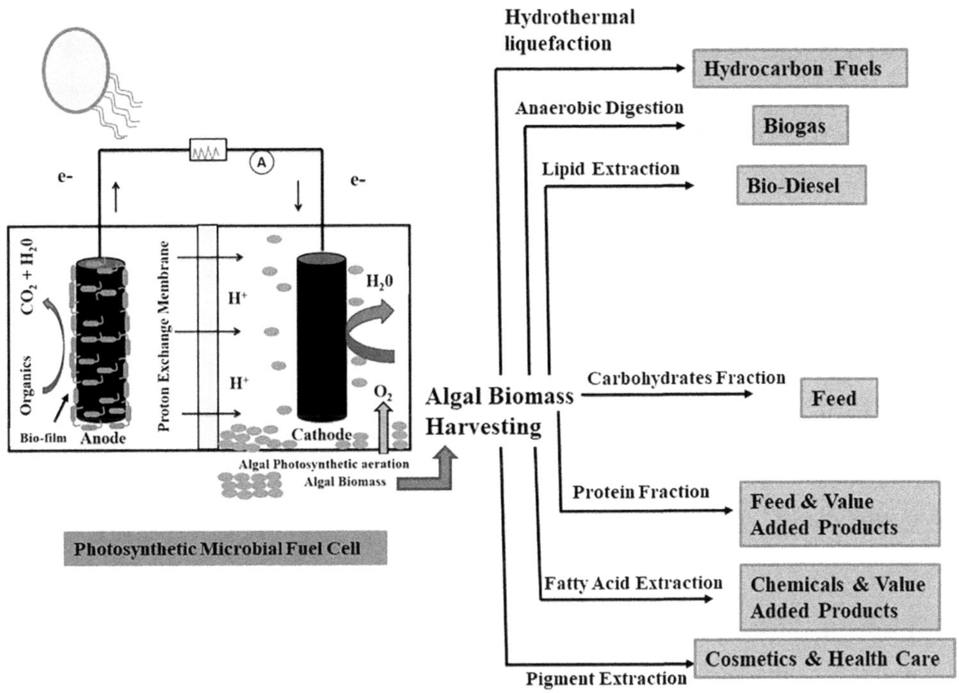

Fig. 4 Biorefinery approach for photosynthetic algae. *(Adopted from S. Arun, A. Sinharoy, K. Pakshirajan, P.N.L. Lens, Algae based microbial fuel cells for wastewater treatment and recovery of value-added products, Renew. Sust. Energ. Rev. 132 (2020), https://doi.org/10.1016/j.rser.2020.110041.)*

autotrophic algae, photosynthesis, and carbon sequestration are achieved due to illumination. The illumination results into productivity in a linear fashion, but after a limit, it adversely affects the performance. The study conducted on the photoperiod and intensity of light had shown that both influence the yield of algal biomass [32,72]. The complex relationship between light intensity and microalgal growth establishes the pattern of direct proportionality between the intensity and algal productivity [32]. In a study, it was observed that the availability of photosynthetic active radiations (PARs), the ratio of light and dark phases, and the continuous or fluctuating light sources affect the photosynthetic rates and thus productivity [31].

4.2 Temperature

The temperature at which the algae is subjected to bloom is very essential for CO_2 biofixation and lipid productivity. The activity of enzyme Rubisco is dependent on the temperature; the low temperature reduces the activity of the enzyme affecting photosynthesis and thus algal biomass production. On the other hand, the extremely high temperature affects algal metabolic activities. The temperature close to room temperature is the best suited for algal activities [32]. The low temperature restricts the MFC performance, as the metabolic reactions are the function of the temperature.

4.3 Flow and mixing

In all types of designs of PMFC, the turbulence and mixing of the three phases are liquid, solids, and gases for the optimum performance of the PMFC. The agitation and mixing avoid agglomeration and sedimentation. The excessive mixing can cause damage to the algal cells, so the optimal mixing speed is desired [32].

4.4 Aeration

Aeration to the cathodic chamber of the microbial fuel cell results in the better generation of electricity and reduction of the protons to water. The aeration in PMFC results into minimum dead space and thus higher productivity. Aeration provides a higher concentration of electron acceptor at the cathode, which in turn provides higher coulombic efficiency.

4.5 Carbon dioxide concentration

The anthropogenic and fossil fuel emissions cause environmental issues like increased CO_2 concentration in the atmosphere leading to global warming, climate change, etc. The algae can sequestrate carbon dioxide into biomass and then to fuel. The carbon dioxide concentration directly affects the function of PMFC. The higher CO_2 concentration lowers the pH and thus hampers the anaerobic digestion of the feed at the anode. While the passage of CO_2 to the cathode chamber will aid the photosynthesis and yield algal biomass [73].

5 Challenges

The efficient CO_2 fixation technology and conversion efficiency are the bottlenecks for the economic feasibility, low productivity, and high cost.

5.1 Design modification

In the last decade, there is a technological advancement in MFC, either as photosynthetic MFC, which leads to the development of PMFC. Further power generation enhancement can be achieved by high surface area electrodes; electrode surface modifications for effective biofilm formation thus transfer the ions. The cost-effective fabrication and the possibility to switch to the continuous type of system will maximize the feed utilization. MFC designs to maximize the utilization of light, CO_2, and nutrients need to be focused [74].

5.2 Obstacles with algal biomass

The harvested algal biomass from the cathodic compartment requires a pretreatment. Alternatively, the algal biomass from the cathodic chamber can be put into the anodic compartment to act as feedstock [67]. They have high reaction velocity, and cell density with resistant to hazardous materials [75].

5.3 Thickness of cathodic biofilm

The power generation is the factor of acceptance of electrons at the cathode, which generates potential difference. The thickness of cathodic biofilm affects the current collection and thus power output of PFMC. As the thickness increases, the power output decreases, and vice versa. Moreover, the biocathodes have a larger startup time [76].

5.4 Temperature and pH

The temperature of the system directly affects the metabolism and thermodynamics of the system and thus oxidation of the feed. The MFC generally operates in moderate temperature conditions. The pH of the anodic chamber is acidic due to the formation of the intermediate acids, while the cathodic pH affects the microalgal growth, which in turn affects biomass productivity. pH plays a role in cell adhesion, morphology, and biocatalytic activity [77,78].

5.5 Feed

The type and composition of carbon source is a matter of concern. The different feed has different productivity. The acetate is one of the most explored feeds used in MFC. Autotrophic biocathodes using inorganic carbon as a source of feed have been widely used in MFCs [79,80].

5.6 Molecular diffusion and cross overs

The molecular diffusion and crossovers of the substances are real-time issues in PMFC. It is the movement of the materials from the anode to the cathode or cathode to the anode through the separating membrane. This results in alteration in potential leading to toxicity or blocking of the active catalytic sites. It also affects the consequent oxygen supply to the cathodic biofilm. The crossover of the material to the anode from the cathode will affect biomass productivity. It overall reduces the coulombic efficiency and performance of the PMFC [81,82].

5.7 Photosynthetic active radiation (PAR)

The closed-type photobioreactors have a facility for artificial illumination, while the open-type reactors are equipped with a natural illumination system, i.e., sunlight. The artificial illumination sometimes lacks photosynthetic active radiation (PAR) which affects the conversion of CO_2 into carbohydrates [83].

Few studies suggest that the biocathode generates low power in comparison to abiotic cathodes, and one of the main challenges is to enhance the power output.

6 Future prospects

It is advantageous to upscale the MFC to PMFC. Algae require minimal nutrients and components; it requires light, CO_2, and H_2O to grow. However, the selection of microalgae species is vital in PMFC performance. This system is cheap, efficient, and high yielding. These are also known to produce biodiesel and other value-added products like carotenoids and other biopolymers from cathodic algal biomass. The PMFC further needs to be studied to explore the other value-added products. The more elaborate studies on to increase the power output of PMFC need to be evaluated.

It has been observed that macroalgae have few advantages over microalgae, especially in terms of nutrient uptake and harvesting. So, it would be worthwhile to explore macroalgae in PMFC. Few of the research are exploring biotechnological and genetically engineering aspects of algae. The slight modification in the cellular structure of algae will result in better power output. Several photobioreactor designs exist for efficient algal growth [84]. Several innovative designs of MFCs can further assist in this. The reactor design is dependent on factors like the organism used, anolyte, catholyte, available light source, and economic considerations as well. Furthermore, mathematical modeling of PMFC systems to validate the experimental with the mathematical model needs to be aimed. It also helps to evaluate the economic and performance based scale-up of the PMFC system. These systems also need to be prepared on the front of provision of the sunlight for better and more efficient algal biomass. In addition, the comparison study on the pure culture and mixed culture needs to be carried out in an effective manner to know more about the efficacy of these cultures. The long-duration MFC needs to be explored for long-term sustainable energy.

7 Conclusion

The advantages of the photosynthetic microbial fuel cells are gaining a lot of focus in top countries like America, China, etc. as well as developing countries like India, and they are trying to excel their vision toward the renewable source of power to humankind. The integration of microalgae with MFC will definitely enhance the efficiency and usefulness of MFC. The algal-based MFC will not only provide a solution for bioenergy generation but also provide a solution for wastewater treatment. Since India is competing with the top developed countries, India should even focus on these photosynthetic organisms, to explore the emerging field of renewable energy.

To improve and scale up the integrated system, more research on the following area needs to be done.

(1) Selection of suitable and efficient algal strains

(2) In-depth study of electron transfer mechanism and CO_2 uptake to improve power generation, biomass control, and recovery of biomass understanding

(3) Scale-up and modeling approach for the algae-based MFCs

(4) Development of integrated systems like MFC-photobioreactor methods to improve biomass recovery and produce value-added products from the algal biomass.

References

[1] P. Saha, B. Paul, Assessment of heavy metal toxicity related with human health risk in the surface water of an industrialized area by a novel technique, Hum. Ecol. Risk. Assess. 25 (2019) 966–987, https://doi.org/10.1080/10807039.2018.1458595.

[2] A. Sarkar, B. Paul, The global menace of arsenic and its conventional remediation—a critical review, Chemosphere 158 (2016) 37–49, https://doi.org/10.1016/j.chemosphere.2016.05.043.

[3] X. Xu, X. Gu, Z. Wang, W. Shatner, Z. Wang, Progress, challenges and solutions of research on photosynthetic carbon sequestration efficiency of microalgae, Renew. Sust. Energ. Rev. 110 (2019) 65–82, https://doi.org/10.1016/j.rser.2019.04.050.

[4] N.V. Mogili, N. Murugesan, S. Ayothiraman, R. Gautam, N.N. Deshavath, R. Reddy Erva, Biohydrogen production from wastewater and organic solid wastes, Waste-to-Energy Approaches Towar Zero Waste (2022) 165–195, https://doi.org/10.1016/B978-0-323-85387-3.00009-4.

[5] R. Gautam, J.K. Nayak, A. Daverey, U.K. Ghosh, Emerging sustainable opportunities for waste to bioenergy: an overview, in: Waste-to-Energy Approaches Towar Zero Waste, Elsevier, 2022, pp. 1–55, https://doi.org/10.1016/B978-0-323-85387-3.00001-X.

[6] Plastic Microplastics in n.d.

[7] R. Gautam, N.V. Rees, R.S. Wilckens, U.K. Ghosh, Hydrogen production in microbial electrolysis cell and reactor digestate valorization for biochar a noble attempt towards circular economy, Int. J. Hydrog. Energy (2023), https://doi.org/10.1016/j.ijhydene.2023.07.190. In press.

[8] A.H. Demirbas, I. Demirbas, Importance of rural bioenergy for developing countries, Energy Convers. Manag. 48 (2007) 2386–2398, https://doi.org/10.1016/J.ENCONMAN.2007.03.005.

[9] K.R. Jegannathan, E.-S. Chan, P. Ravindra, Harnessing biofuels: a global renaissance in energy production? Renew. Sust. Energ. Rev. 13 (2009) 2163–2168, https://doi.org/10.1016/J.RSER.2009.01.012.

[10] R. Gautam, J.K. Nayak, K.N. Talapatra, Amit, U.K. Ghosh, Assessment of different organic substrates for bio-electricity and bio-hydrogen generation in an integrated bio-electrochemical system, Mater. Today Proc. (2021) 6–10, https://doi.org/10.1016/j.matpr.2021.06.223.

[11] N. Talapatra, R. Gautam, V. Mittal, U.K. Ghosh, A comparative study of the growth of microalgae-bacteria symbiotic consortium with the axenic culture of microalgae in dairy wastewater through extraction and quantification of chlorophyll, Mater. Today Proc. (2021), https://doi.org/10.1016/j.matpr.2021.06.227.

[12] Amit, J.K. Nayak, U.K. Ghosh, Microalgal remediation of anaerobic pretreated pharmaceutical wastewater for sustainable biodiesel production and electricity generation, J. Water Process Eng. 35 (2020) 101192, https://doi.org/10.1016/j.jwpe.2020.101192.

[13] J.K. Nayak, Amit, U.K. Ghosh, An innovative mixotrophic approach of distillery spent wash with sewage wastewater for biodegradation and bioelectricity generation using microbial fuel cell, J. Water Process Eng. 23 (2018) 306–313, https://doi.org/10.1016/j.jwpe.2018.04.003.

[14] Amit, R. Chandra, U.K. Ghosh, J.K. Nayak, Phycoremediation potential of marine microalga Tetraselmis indica on secondary treated domestic sewage for nutrient removal and biodiesel production, Environ. Sci. Pollut. Res. 24 (2017) 20868–20875, https://doi.org/10.1007/s11356-017-9734-6.

[15] I. Nasser, A. Sulaimi, J.K. Nayak, H. Alhimali, A. Sana, A. Al Mamun, Effect of volatile fatty acids accumulation on biogas production by sludge-feeding thermophilic anaerobic digester and predicting process parameters, Fermentation (2022), https://doi.org/10.3390/fermentation8040184.
[16] R. Gautam, J.K. Nayak, N.V. Rees, R.S. Wilckens, U.K. Ghosh, Bio-hydrogen production through microbial electrolysis cell: structural components and influencing factors, Chem. Eng. J. 455 (2023), https://doi.org/10.1016/j.cej.2022.140535.
[17] J.K. Nayak, R. Gautam, U.K. Ghosh, Bioremediation potential of bacterial consortium on different wastewaters for electricity and biomass feedstock generation, Biomass Conv. Bioref. (2022), https://doi.org/10.1007/s13399-022-02992-2.
[18] Amit, N. Kumar, S. Verma, J. Park, A.K. Jaiswal, U.K. Ghosh, R. Gautam, Utilization of nano-sized waste lime sludge particles in harvesting marine microalgae for biodiesel feedstock production, Nanotechnol. Environ. Engi. (2022), https://doi.org/10.1007/s41204-021-00195-0.
[19] R. Gautam, V. Mittal, A. Chauhan, A. Jaiswal, U.K. Ghosh, Biochar-derived green catalysts for biofuel production, in: Encyclopedia of Green Materials, Springer, 2022.
[20] V. MIttal, K.N. Talapatra, R. Gautam, U.K. Ghosh, Nanocatalyst for production of biodiesel, in: Encyclopedia of Green Materials, Springer, 2023.
[21] A. Maiti, A. Pandey, Polymer and waste plastic in membranes, in: Reference Module in Materials Science and Materials Engineering, Elsevier, 2022.
[22] B. Abdullah, S.A.F. Syed Muhammad, Z. Shokravi, S. Ismail, K.A. Kassim, A.N. Mahmood, et al., Fourth generation biofuel: a review on risks and mitigation strategies, Renew. Sust. Energ. Rev. 107 (2019) 37–50, https://doi.org/10.1016/j.rser.2019.02.018.
[23] S.S. Kumar, V. Kumar, R. Kumar, S.K. Malyan, N.R. Bishnoi, Ferrous sulfate as an in-situ anodic coagulant for enhanced bioelectricity generation and COD removal from landfill leachate, Energy 176 (2019) 570–581.
[24] C. Santoro, C. Arbizzani, B. Erable, I. Ieropoulos, Microbial fuel cells: from fundamentals to applications. A review, J. Power Sources 356 (2017) 225–244, https://doi.org/10.1016/j.jpowsour.2017.03.109.
[25] M.K. Enamala, R. Dixit, A. Tangellapally, M. Singh, S.M.P. Dinakarrao, M. Chavali, et al., Photosynthetic microorganisms (algae) mediated bioelectricity generation in microbial fuel cell: concise review, Environ. Technol. Innov. 19 (2020) 100959, https://doi.org/10.1016/j.eti.2020.100959.
[26] W. Apollon, A.I. Luna-Maldonado, S.K. Kamaraj, J.A. Vidales-Contreras, H. Rodríguez-Fuentes, J.F. Gómez-Leyva, et al., Progress and recent trends in photosynthetic assisted microbial fuel cells: a review, Biomass Bioenergy (2021) 148, https://doi.org/10.1016/j.biombioe.2021.106028.
[27] R.G. Smith, I.J. Smith, B.D. Smith, A novel strategy for sequestering atmospheric CO_2: the use of sealed microalgal cultures located in the open-oceans, Renew. Sust. Energ. Rev. 83 (2018) 85–89, https://doi.org/10.1016/j.rser.2017.10.001.
[28] P. Chen, M. Min, Y. Chen, L. Wang, Y. Li, Q. Chen, et al., Review of the biological and engineering aspects of algae to fuels approach, Int. J. Agric. Biol. Eng. 2 (2009) 1–30, https://doi.org/10.3965/j.issn.1934-6344.2009.04.001-030.
[29] N. Seyed Hosseini, H. Shang, J.A. Scott, Biosequestration of industrial off-gas CO2 for enhanced lipid productivity in open microalgae cultivation systems, Renew. Sust. Energ. Rev. 92 (2018) 458–469, https://doi.org/10.1016/j.rser.2018.04.086.
[30] M. El-Dairi, R.J. House, Optic nerve hypoplasia, in: Handbook of Pediatric Retinal OCT and the Eye-Brain Connection, Elsevier, 2019, pp. 285–287, https://doi.org/10.1016/B978-0-323-60984-5.00062-7.
[31] J.U. Grobbelaar, Microalgal biomass production: challenges and realities, Photosynth. Res. 106 (2010) 135–144, https://doi.org/10.1007/s11120-010-9573-5.
[32] B. Zhao, Y. Su, Process effect of microalgal-carbon dioxide fixation and biomass production: a review, Renew. Sust. Energ. Rev. 31 (2014) 121–132, https://doi.org/10.1016/j.rser.2013.11.054.
[33] E. Sforza, P. Kumkum, E. Barbera, S. Kumar, Bioremediation of industrial effluents: how a biochar pretreatment may increase the microalgal growth in tannery wastewater, J. Water Process Eng. 37 (2020) 101431, https://doi.org/10.1016/j.jwpe.2020.101431.
[34] G. Akerman-Sanchez, K. Rojas-Jimenez, Fungi for the bioremediation of pharmaceutical-derived pollutants: a bioengineering approach to water treatment, Environ. Adv. 4 (2021) 100071, https://doi.org/10.1016/j.envadv.2021.100071.

[35] M. Nanda, B. Chand, S. Kharayat, T. Bisht, N. Nautiyal, S. Deshwal, et al., Integration of microalgal bioremediation and biofuel production: a 'clean up' strategy with potential for sustainable energy resources, Curr Res Green Sustain Chem 4 (2021) 100128, https://doi.org/10.1016/j.crgsc.2021.100128.

[36] S. Gupta, P. Srivastava, S.A. Patil, A.K. Yadav, A comprehensive review on emerging constructed wetland coupled microbial fuel cell technology: potential applications and challenges, Bioresour. Technol. (2020) 124376, https://doi.org/10.1016/j.biortech.2020.124376.

[37] B.E. Logan, J.M. Regan, Microbial fuel cells—challenges and applications, Environ. Sci. Technol. 40 (2006) 5172–5180, https://doi.org/10.1021/es0627592.

[38] D. Pant, G. Van Bogaert, L. Diels, K. Vanbroekhoven, A review of the substrates used in microbial fuel cells (MFCs) for sustainable energy production, Bioresour. Technol. 101 (2010) 1533–1543, https://doi.org/10.1016/j.biortech.2009.10.017.

[39] Y. Fan, H. Liu, Materials for microbial fuel cells, in: Materials for Low-Temperature Fuel Cells, John Wiley & Sons, 2015, pp. 145–166, https://doi.org/10.1002/9783527644308.ch07.

[40] M.F. Umar, S.Z. Abbas, M.N. Mohamad Ibrahim, N. Ismail, M. Rafatullah, Insights into advancements and electrons transfer mechanisms of electrogens in benthic microbial fuel cells, Membranes 10 (2020) 1–18, https://doi.org/10.3390/membranes10090205.

[41] Let's talk science, Microbial Fuel Cells, 2019. https://letstalkscience.ca/educational-resources/stem-explained/microbial-fuel-cells. (Accessed 14 August 2023).

[42] M. Rahimnejad, A. Adhami, S. Darvari, A. Zirepour, S.E. Oh, Microbial fuel cell as new technology for bioelectricity generation: a review, Alexandria Eng J 54 (2015) 745–756, https://doi.org/10.1016/j.aej.2015.03.031.

[43] D. Pant, G. Van Bogaert, L. Diels, K. Vanbroekhoven, Bioresource technology a review of the substrates used in microbial fuel cells (MFCs) for sustainable energy production, Bioresour. Technol. (2009), https://doi.org/10.1016/j.biortech.2009.10.017.

[44] R.K. Jung, S. Cheng, S.E. Oh, B.E. Logan, Power generation using different cation, anion, and ultrafiltration membranes in microbial fuel cells, Environ. Sci. Technol. 41 (2007) 1004–1009, https://doi.org/10.1021/es062202m.

[45] P. Aelterman, K. Rabaey, P. Clauwaert, W. Verstraete, Microbial fuel cells for wastewater treatment, Water Sci. Technol. 54 (2006) 9–15, https://doi.org/10.2166/wst.2006.702.

[46] M. Behera, P.S. Jana, M.M. Ghangrekar, Performance evaluation of low cost microbial fuel cell fabricated using earthen pot with biotic and abiotic cathode, Bioresour. Technol. 101 (2010) 1183–1189, https://doi.org/10.1016/j.biortech.2009.07.089.

[47] S.P. Jung, M.H. Yoon, S.M. Lee, S.E. Oh, H. Kang, J.K. Yang, Power generation and anode bacterial community compositions of sediment fuel cells differing in anode materials and carbon sources, Int. J. Electrochem. Sci. 9 (2014) 315–326.

[48] S. Jung, J.M. Regan, Comparison of anode bacterial communities and performance in microbial fuel cells with different electron donors, Appl. Microbiol. Biotechnol. 77 (2007) 393–402, https://doi.org/10.1007/s00253-007-1162-y.

[49] A.P. Kruzic, S.K. Liehr, Natural treatment and onsite systems, Water Environ. Res. 80 (2008) 1206–1224, https://doi.org/10.2175/106143008x328608.

[50] D.F. Juang, C.H. Lee, S.C. Hsueh, Comparison of electrogenic capabilities of microbial fuel cell with different light power on algae grown cathode, Bioresour. Technol. 123 (2012) 23–29, https://doi.org/10.1016/j.biortech.2012.07.041.

[51] S.S. Kumar, S.K. Malyan, A. Kumar, S. Basu, N.R. Bishnoi, Microbial fuel cells technology: food to energy conversion by anode respiring bacteria, in: Environmental Concerns of 21st Century: Indian and Global Context, 2016, pp. 139–154.

[52] I. Gajda, J. Greenman, C. Melhuish, I. Ieropoulos, Self-sustainable electricity production from algae grown in a microbial fuel cell system, Biomass Bioenergy 82 (2015) 87–93, https://doi.org/10.1016/j.biombioe.2015.05.017.

[53] M. Zhou, H. He, T. Jin, H. Wang, Power generation enhancement in novel microbial carbon capture cells with immobilized *Chlorella vulgaris*, J. Power Sources 214 (2012) 216–219, https://doi.org/10.1016/j.jpowsour.2012.04.043.

[54] E.E. Powell, M.L. Mapiour, R.W. Evitts, G.A. Hill, Growth kinetics of Chlorella vulgaris and its use as a cathodic half cell, Bioresour. Technol. 100 (2009) 269–274, https://doi.org/10.1016/j.biortech.2008.05.032.

[55] A. González Del Campo, P. Cañizares, M.A. Rodrigo, F.J. Fernández, J. Lobato, Microbial fuel cell with an algae-assisted cathode: a preliminary assessment, J. Power Sources 242 (2013) 638–645, https://doi.org/10.1016/j.jpowsour.2013.05.110.

[56] X. Wang, Y. Feng, J. Liu, H. Lee, C. Li, N. Li, et al., Sequestration of CO2 discharged from anode by algal cathode in microbial carbon capture cells (MCCs), Biosens. Bioelectron. 25 (2010) 2639–2643, https://doi.org/10.1016/j.bios.2010.04.036.

[57] C.C. Lin, C.H. Wei, C.I. Chen, C.J. Shieh, Y.C. Liu, Characteristics of the photosynthesis microbial fuel cell with a Spirulina platensis biofilm, Bioresour. Technol. 135 (2013) 640–643, https://doi.org/10.1016/j.biortech.2012.09.138.

[58] C.C. Fu, T.C. Hung, W.T. Wu, T.C. Wen, C.H. Su, Current and voltage responses in instant photosynthetic microbial cells with Spirulina platensis, Biochem. Eng. J. 52 (2010) 175–180, https://doi.org/10.1016/j.bej.2010.08.004.

[59] S.K. Malyan, S. Singh, A. Bachheti, M. Chahar, M.K. Sah, A. Kumar, A.N. Yadav, S.S. Kumar, Cyanobacteria: a perspective paradigm for agriculture and environment, in: New and Future Developments in Microbial Biotechnology and Bioengineering, Elsevier, 2020, pp. 215–224.

[60] D.J. Farrelly, C.D. Everard, C.C. Fagan, K.P. McDonnell, Carbon sequestration and the role of biological carbon mitigation: a review, Renew. Sust. Energ. Rev. 21 (2013) 712–727, https://doi.org/10.1016/j.rser.2012.12.038.

[61] R. Lal, Carbon sequestration, Philos. Trans. R. Soc. B 363 (2008) 815–830, https://doi.org/10.1098/rstb.2007.2185.

[62] W. Yun-Hai, W. Bai-Shi, P. Bin, C. Qing-Yun, Y. Wei, Electricity production from a bio-electrochemical cell for silver recovery in alkaline media, Appl. Energy 112 (2013) 1337–1341, https://doi.org/10.1016/j.apenergy.2013.01.012.

[63] J.K. Nayak, U.K. Ghosh, Microalgae cultivation for pretreatment of pharmaceutical wastewater associated with microbial fuel cell and biomass feed stock production, Adv. Sci. Technol. Innov. (2020) 383–387, https://doi.org/10.1007/978-3-030-13068-8_96.

[64] H. Salehizadeh, N. Yan, R. Farnood, Recent advances in microbial CO_2 fixation and conversion to value-added products, Chem. Eng. J. 390 (2020) 124584, https://doi.org/10.1016/j.cej.2020.124584.

[65] R. Gayathri, S. Mahboob, M. Govindarajan, K.A. Al-Ghanim, Z. Ahmed, N. Al-Mulhm, et al., A review on biological carbon sequestration: a sustainable solution for a cleaner air environment, less pollution and lower health risks, J. King Saud Univ. Sci. (2021) 33, https://doi.org/10.1016/j.jksus.2020.101282.

[66] M.A. Abdelkareem, M.A. Lootah, E.T. Sayed, T. Wilberforce, H. Alawadhi, B.A.A. Yousef, et al., Fuel cells for carbon capture applications, Sci. Total Environ. 769 (2021) 144243, https://doi.org/10.1016/j.scitotenv.2020.144243.

[67] S. Arun, A. Sinharoy, K. Pakshirajan, P.N.L. Lens, Algae based microbial fuel cells for wastewater treatment and recovery of value-added products, Renew. Sust. Energ. Rev. (2020) 132, https://doi.org/10.1016/j.rser.2020.110041.

[68] Amit, U. Kumar Ghosh, Utilization of kinnow peel extract with different wastewaters for cultivation of microalgae for potential biodiesel production, J. Environ. Chem. Eng. 7 (2019) 103135, https://doi.org/10.1016/j.jece.2019.103135.

[69] J.K. Nayak, U.K. Ghosh, Post treatment of microalgae treated pharmaceutical wastewater in photosynthetic microbial fuel cell (PMFC) and biodiesel production, Biomass Bioenergy (2019) 131, https://doi.org/10.1016/j.biombioe.2019.105415.

[70] M. Li, M. Zhou, X. Tian, C. Tan, T. Gu, Enhanced bioenergy recovery and nutrient removal from swine wastewater using an airlift-type photosynthetic microbial fuel cell, Energy 226 (2021) 120422, https://doi.org/10.1016/j.energy.2021.120422.

[71] Y.K. Leong, K.W. Chew, W.H. Chen, J.S. Chang, P.L. Show, Reuniting the biogeochemistry of algae for a low-carbon circular bioeconomy, Trends Plant Sci. (2021) 1–12, https://doi.org/10.1016/j.tplants.2020.12.010.

[72] Y. Li, W. Zhou, B. Hu, M. Min, P. Chen, R.R. Ruan, Effect of light intensity on algal biomass accumulation and biodiesel production for mixotrophic strains Chlorella kessleri and Chlorella protothecoide cultivated in highly concentrated municipal wastewater, Biotechnol. Bioeng. 109 (2012) 2222–2229, https://doi.org/10.1002/bit.24491.

[73] S.P. Singh, P. Singh, Effect of CO_2 concentration on algal growth: a review, Renew. Sust. Energ. Rev. 38 (2014) 172–179, https://doi.org/10.1016/j.rser.2014.05.043.

[74] A. Elmekawy, H.M. Hegab, K. Vanbroekhoven, D. Pant, Techno-productive potential of photosynthetic microbial fuel cells through different configurations, Renew. Sust. Energ. Rev. 39 (2014) 617–627, https://doi.org/10.1016/j.rser.2014.07.116.

[75] W. Klinthong, Y.H. Yang, C.H. Huang, C.S. Tan, A review: microalgae and their applications in CO_2 capture and renewable energy, Aerosol Air Qual. Res. 15 (2015) 712–742, https://doi.org/10.4209/aaqr.2014.11.0299.

[76] A.E. Franks, N. Malvankar, K.P. Nevin, Bacterial biofilms: the powerhouse of a microbial fuel cell, Biofuels 1 (2010) 589–604, https://doi.org/10.4155/bfs.10.25.

[77] J. Ling, Y. Xu, C. Lu, W. Lai, G. Xie, et al., Enhancing stability of microalgae biocathode by a partially submerged carbon cloth electrode for bioenergy production from wastewater, Energies 12 (2019) 3229.

[78] A.W. Almutairi, A.E.K.B. El-Sayed, M.M. Reda, Combined effect of salinity and pH on lipid content and fatty acid composition of Tisochrysis lutea, Saudi J Biol Sci 27 (2020) 3553–3558, https://doi.org/10.1016/j.sjbs.2020.07.027.

[79] D. Massazza, A.J. Robledo, C.N. Rodriguez Simón, J.P. Busalmen, S. Bonanni, Energetics, electron uptake mechanisms and limitations of electroautotrophs growing on biocathodes—a review, Bioresour. Technol. (2021) 342, https://doi.org/10.1016/j.biortech.2021.125893.

[80] D. Ucar, Y. Zhang, I. Angelidaki, An overview of electron acceptors in microbial fuel cells, Front. Microbiol. 8 (2017) 1–14, https://doi.org/10.3389/fmicb.2017.00643.

[81] V. Chaturvedi, P. Verma, Microbial fuel cell: a green approach for the utilization of waste for the generation of bioelectricity, Bioresour. Bioprocess. (2016) 3, https://doi.org/10.1186/s40643-016-0116-6.

[82] X.C. Quan, Y.P. Quan, K. Tao, Effect of anode aeration on the performance and microbial community of an air-cathode microbial fuel cell, Chem. Eng. J. 210 (2012) 150–156, https://doi.org/10.1016/j.cej.2012.09.009.

[83] Q. Huang, F. Jiang, L. Wang, C. Yang, Design of photobioreactors for mass cultivation of photosynthetic organisms, Engineering 3 (2017) 318–329, https://doi.org/10.1016/J.ENG.2017.03.020.

[84] Y. Duan, F. Shi, Bioreactor Design for Algal Growth as a Sustainable Energy Source, Elsevier B.V., 2014, https://doi.org/10.1016/B978-0-444-59566-9.00002-8.

CHAPTER 7

Environmental applications of bioelectrochemical fuel cells

Atul Kumar[a], Supriya Sehrawat[a], Priya[a], Aryan Boora[a], Bhavna[a], Parmod Kumar[b], Suresh Kumar[c], and Surender Duhan[a]

[a]Advanced Sensors Lab, Department of Physics, Deenbandhu Chhotu Ram University of Science and Technology, Sonipat, Haryana, India

[b]Department of Physics, JC Bose University of Science and Technology, YMCA, Faridabad, Haryana, India

[c]Department of Electronics, Kurukshetra University, Kurukshetra, India

1 Introduction

The development of analytical techniques that allow quick "on-the-spot" performances for medical diagnostics and measuring environmental contaminants has attracted much attention in recent years. A device used in distant locations, such as an ambulance, an agricultural farm, or field monitoring of environmental contaminants, should be able to give quick, sensitive, compact, and low-cost measurements. Furthermore, for field measurements, the equipment should be straightforward to use by a nonqualified individual [1].

Bioelectrochemical sensors convert biological data, such as analyte concentrations, to an evaluatable signal, such as current or voltage [2]. However, this kind of sensor does not need a separate transducer since the analyte concentration is determined immediately from the recorded electrical response proportionate to the analyte concentration received from the electrode [3]. Bioelectrochemical sensors rely on a catalytic interaction between immobilized biomolecules and the analyte of interest which generates or consumes electrons or ions by altering the solution's electrical characteristics, such as its electric potential or current [4]. Electrochemical sensing requires three electrodes: a reference electrode, a working electrode, and a counter electrode. Both a biomolecular identification system and a physiochemical transducer serve as the transduction element, which is also referred to as the working electrode, redox, or sensing electrode. Ag/AgCl is used as the reference electrode [5]. To maintain a constant voltage, the reference electrode is maintained away from the reaction site. By connecting the counter or auxiliary electrode electrolytic solution to the working electrode, current may be transferred to the working electrode. Due to their conductivity and chemical stability, gold, platinum, carbon, and silicon electrodes are often employed [6].

In bioelectrochemical sensors, amperometric, potentiometric, and conductometric transducers transform the response of bioreceptor molecules to measurable signals [7].

Algae Based Bioelectrochemical Systems for Carbon Sequestration, Carbon Storage, Bioremediation and Bioproduct Generation
https://doi.org/10.1016/B978-0-323-91023-1.00008-9

"As the name suggests, amperometric transducers depend on changes in electric current. Current flow fluctuations are caused by electrochemical redox reactions. This change in current flow is indicative of a contact between the analyte and the bioreceptor, indicating the bioreceptor molecule's reaction to the analyte, which is restricted by the analyte's mass shift to the electrode. Potentiometric transducers, on the other hand, detect charge aggregation in an electrochemical cell composed of an ion-selective electrode that interacts with charged ions. As a consequence of this connection, the potential charge congregates at the reference electrode. The charge potential is exactly proportional to the logarithm of the analyte concentration, according to the Nernst equation. Impedance transducers detect changes in resistance and capacitance, which are often induced by a biorecognition event. During the biorecognition process, a three-tier electrode is sufficiently selective to allow the biological analyte to immobilize on the bioreceptor's surface. It is accomplished via the application of voltage and subsequent measurement of current flow. The impedance analyzer monitors and controls the change in impedance. Conductometric transducers are distinguished from other types of transducers by the fact that they monitor changes in the conductive properties of the reaction medium. Biorecognition modifies the ionic concentration, which modifies the reaction media, resulting in changes in conductive flow" [3].

Bioelectrochemical sensors aided in the development of commercial point-of-care systems by enabling the use of glucose, uric acid, and cholesterol detectors as diagnostic tools [8]. These technologies have risen in popularity as concerns about water quality monitoring and pollution management have increased, with portable sensors that are real time, simple, high sensitivity, and quick in their analysis becoming more common [9]. Nanomaterial and microfabrication methods for electrode surface modification will be required for future advancements in electrochemical sensing platforms. Carbon nanotubes and conductive polymers are now being used to improve sensitivity, selectivity, stability, and long-term storage [10]. On the other hand, nonspecific adsorption continues to be a stumbling barrier for any biosensors designed to deal with "real" samples. Additionally, it is important to recognize the limitations of today's technology. Separation costs, time-consuming preparation, and possible toxicity are only some of the problems that remain unresolved in the creation of carbon nanotubes. Additionally, it has been said that CNT's present electrical characteristics fall short of their theoretical potential [11]. Similarly, despite considerable progress over the last three decades, there is still a general lack of robustness; nevertheless, viable solutions to some of these issues have already appeared, and others are anticipated to come shortly.

2 Systems for bioelectrochemical sensing

Bioelectrochemical sensing systems synthesize energy, hydrogen, or other useful molecules by combining biological and electrochemical processes. These are two of the most prominent instances of rapidly expanding biotechnology, referred to as microbial fuel cells (MFCs) and microbial electrolysis cells (MECs) [12].

2.1 Microbial fuel cells

A "microbial fuel cell (MFC) is a device that uses microorganisms to convert chemical energy to electrical energy" [13]. Bioanodes and biocathodes are used to construct these electrochemical cells. A membrane separates the anode and cathode chambers in the majority of MFCs. The electrons produced during oxidation are transported to an electrode or to a redox mediator species in a very short period of time [14]. This is accomplished by transmitting electron flow from the cathode to the anode. The charge balance of the system is maintained via ionic transport inside the cell, which occurs most often across an ionic membrane. In order to create CO_2, protons, and electrons, the majority of MFCs depend on an organic electron donor, which is oxidized to produce CO_2 [1,15]. Various other electron donors, including sulfur compounds and hydrogen, have also been identified. The cathode process makes use of a variety of electron acceptors, with oxygen being the most often seen. Additional studies have looked at metal recovery via reduction, water to hydrogen, nitrate reduction, and sulfate reduction, among others [16].

2.1.1 Efficacy mechanism

When aerobic bacteria eat a material such as sugar, they generate carbon dioxide and water. When there is no oxygen present, they produce carbon dioxide, hydrogen ions, and electrons as byproducts. Inorganic mediators are used by microbial fuel cells to access the electron transport chain [16,17]. The mediator transports electrons from the electron transport chain to the bacterial outer membrane, where they would have been taken up by oxygen or other intermediates. When the mediator runs out of electrons, it transfers them to an electrode, becoming the anode. After the electrons are liberated, the mediator is oxidized and ready to proceed [18]. If oxygen is present, its greater electronegativity traps the electrons. Bacteria in the anodic chamber identify the anode as the terminal electron acceptor. As a consequence, the anode's redox potential influences microbial activity. Anodic potential appears to be the key to maximum power generation. Mediators include natural red, methylene blue, thionine, and resorufin. Exoelectrogens are creatures that can generate electricity. To utilize this current, exoelectrogens must be confined in a fuel cell [19].

The mediator is mixed with a microorganism, such as yeast, in a glucose-containing solution. Because the bacteria are trapped in a chamber, they must utilize anaerobic respiration. An electrode is used as the anode [20]. The MFC's second chamber holds solution in addition to the cathode. The oxygen sink at the end of the electron transport chain is the biological cell. An oxidizing substance in solution absorbs electrons at the cathode. A solid oxidizing agent is used owing to its smaller size than the electron chain in the yeast cell. Oxygen or a solid oxidizing agent provides most of the chemical energy in the cell. A cable or other electrically conductive route connects the two electrodes. A salt bridge or an ion-exchange membrane connects the two chambers to complete the circuit. This last feature allows proton generation.

The reduced mediator transports electrons from the cell to the electrode. It oxidizes as it deposits electrons. By conducting these charges to an electron sink, a wire is created. Then they are oxidized. A proton-exchange membrane such as Nafion also transports hydrogen ions/protons between the anode and cathode. They will combine with oxygen despite the smaller concentration gradient, but they will require an electron. It generates power and utilizes hydrogen to maintain a gradient. When used as a substrate in a microbial fuel cell, algal biomass generates a lot of energy [21].

2.1.2 Different types of microbial fuel cells

2.1.2.1 Mediated

Electrochemically, "the majority of microbial cells are inactive. Mediators such as thionine, methyl viologen, methyl blue, humic acid, and neutral red assist in electron transport from microbial cells to the electrode. The vast majority of today's mediators are both costly and hazardous" [22].

2.1.2.2 Without a mediator

Electrochemically active bacteria transport electrons to the electrode in mediator-free microbial fuel cells (electrons are carried directly from the bacterial respiratory enzyme). Electrochemically active bacteria include *Shewanella putrefaciens* and *Aeromonas hydrophila.* Certain bacteria use pili to transport electrons from their electron producing cells to their outer membrane. The kind of ion-exchange membrane utilized in mediator-free MFCs and the system properties are less well understood.

Without the need of a mediator, microbial fuel cells may function on wastewater and get energy directly from plants and O_2. This is referred to as a plant microbial fuel cell system. Reed sweet grass, cordgrass, rice, tomatoes, lupines, and algae are some of the plants that are utilized. Due to the fact that this version is powered by living plants, it has the potential to be eco-friendly [23].

2.1.2.3 Electrolysis of bacteria

Microbial electrolysis cells are one kind of mediator-free MFC (MEC). "MECs generate hydrogen or methane in part by providing a voltage to bacteria, while MFCs generate an electric current via bacterial decomposition of organic molecules in water. It boosts the voltage produced by microbial breakdown of organic matter, resulting in water electrolysis or methane generation. Microbial electrosynthesis is the polar opposite of MFC, in which bacteria utilize an external electric current to decrease carbon dioxide and create multicarbon organic molecules" [24].

2.1.2.4 Terrestrial

Soil-based microbial fuel cells are built on the fundamental principles of MFCs, with soil serving as the anodic medium, inoculum, and proton-exchange membrane. The cathode is exposed to the air and sits on top of the anode, which is buried in the soil to a specified depth.

Soils are home to a wide array of microorganisms, including the electrogenic bacteria required for the production of MFCs, as well as complex sugars and other nutrients obtained from plant and animal breakdown. Additionally, the aerobic (oxygen-consuming) microorganisms in the soil serve as an oxygen filter, similar to the costly proton-exchange membrane materials employed in laboratory MFC systems, resulting in a reduction in the soil's redox potential as depth increases. MFCs made of soil are gaining popularity in science classrooms.

Sediment microbial fuel cells (SMFCs) have been used to cleanse wastewater. Simple solid-state fuel cells (SMFCs) have the ability to generate electricity while also removing contaminants from wastewater. The majority of these SMFCs replicate artificial wetlands via the use of plants. By 2015, testing of SMFCs has increased to over 150L [25].

In 2015, scientists claimed that they had developed an SMFC application capable of collecting energy and charging a battery. Salts dissolve in water, producing positively and negatively charged ions that move and attach to the negative and positive electrodes, thus charging the battery and allowing the salt to be evacuated, enabling microbial capacitive desalination. Microorganisms provide energy in excess of what the desalination process requires. Additionally, the biological process that produces energy purifies leftover water before to its release into the environment or reuse in agricultural and industrial applications. Aqualia accomplished this via the opening of its desalination innovation center in Denia, Spain, in early 2020.

2.1.2.5 Biofilm photosynthesis

A phototrophic biofilm is one that survives only on light. MFCs make use of phototrophic biofilm anodes composed of photosynthetic bacteria such as Chlorophyta and Cyanophyta. As a consequence of photosynthesis, organic molecules are formed and electrons are donated. According to one study, PBMFCs have a high enough power density to be used in practical applications. Biological photovoltaic systems are a kind of phototrophic MFC in which the anode is made entirely of oxygenic photosynthetic material [26].

2.1.2.6 Membrane nanoporeous

The United States Naval Research Laboratory created nanoporous membrane microbial fuel cells that use a non-PEM to provide passive diffusion inside the cell. The membrane is a polymer filter that is impervious to water (nylon, cellulose, or polycarbonate). While it has similar power densities as Nafion (a well-known PEM), it is more durable. Passive diffusion is enabled by porous membranes, which reduces the amount of energy required to maintain the PEM active and therefore increases overall energy output [27].

MFCs without a membrane may be used to introduce anaerobic bacteria into aerobic settings. On the other hand, membrane-free MFCs are contaminated at the cathode by indigenous bacteria and the power supply microorganism. To achieve the benefits of a membrane-less MFC, a new passive diffusion of nanoporous membranes without risk of cathode contamination may be utilized.

2.1.2.7 Ceramic membrane

Ceramic materials may be used in lieu of PEM membranes. Costs for a ceramic membrane may be as low as $5.66 per square meter. The macroporous structure of ceramic membranes enables efficient transport of ionic species. Earthenware, alumina, mullite, pyrophyllite, and terracotta are just a few of the materials that have been effectively used in ceramic MFCs.

2.2 Microbial electrolysis cell

"Microbial electrolysis cells (MECs) are microbial fuel cells that generate electricity via the usage of bacteria (MFC)" [28]. MECs reverse the process of producing hydrogen or methane from organic material in part by using an electric current. Simultaneously, MFCs generate electricity via the microbial decomposition of organic molecules. In an ideal world, electric current would be generated by a renewable energy source. The hydrogen or methane generated may be used to power a second PEM fuel cell or a conventional internal combustion engine.

MEC systems are composed of the following components:

- Microorganisms—microorganisms are organisms that live in close proximity to the anode. Microorganism identification dictates the MEC's products and efficiency.
- Carbon fabric, carbon paper, graphite felt, graphite granules, or graphite brushes are all acceptable anode materials in a MEC, just as they are in an MFC. Platinum may be used as a catalyst to decrease the amount of surplus potential needed to produce hydrogen. Due to the exorbitant expense of platinum, researchers are exploring the possibility of using biocathodes instead. Stainless steel plates were utilized as the cathode and anode materials in place of catalysts. Additionally, membranes (although some MECs are membranes), tubing, and gas-collecting systems are included.

2.2.1 Efficacy mechanism

MEC is composed of three main components: anode chamber, cathode chamber, and separator. In a MEC, electrochemically active microorganisms growing on the anode's surface decompose organic molecules into CO_2, electrons, and protons. Electrons and protons travel to the cathode, where they combine to create hydrogen, through the electrolyte and the external electric circuit, respectively. A microbial biofilm contributes to the oxidation of the anode on the electrode acting as an electrocatalyst. MEC intends to develop novel microbial biofilms in order to produce MEC anodes. This will be accomplished by identifying new sources of inoculum, changing the microbial population, and defining the form of MEC anodes [29].

3 Bioelectrochemical sensing applications

To broaden the use of bioelectrochemical sensing, researchers have concentrated on material selection and optimization, device design, and screening of electrochemically active or inert model microorganisms. Notably, electron transport and energy conversion, or the exchange of chemical and electrical energy, are central to these applications and research [30].

3.1 Treatment of waste water

Untreated wastewater discharge into the environment has serious consequences for ecosystems and public health [31]. Traditional wastewater treatment techniques are costly to implement and provide a poor rate of return on investment. Bioelectrochemical sensing devices seem to be a viable alternative to traditional techniques in this regard. Microbial fuel cells and microbial electrolysis cells are the most extensively researched bioelectrochemical sensing devices for wastewater treatment and concurrent product recovery. Microbial fuel cells are primarily used to generate energy and to recover/remove nutrients from the environment (nitrogen and phosphorus) [32,33].

In comparison, microbial electrolysis cells are primarily used to extract biofuels from wastewater (ethanol, methane, and hydrogen). The majority of investigations have used laboratory-scale bioelectrochemical sensing devices and have concluded that further in-depth study is necessary. As a result, the feasibility of these systems on a broad scale continues to be a major barrier to general implementation [33].

3.2 Generation of electricity

Microbial fuel cells and microbial electrolysis cells are two types of bioelectrochemical sensing systems that use a biofilm on an electrode as a biocatalyst to convert biodegradable organic matter to electrical energy or hydrogen. Recently, waste-to-energy technology based on microbial fuel cells [34] has been used to remediate organic contaminants in household or industrial wastewater while also producing power. Maximum power density depending on reactor capacity raised to 1kW per cubic meter. By substituting energy-intensive aeration with bioelectrochemical systems that generate electricity from waste, bioelectrochemical systems have the potential to significantly decrease energy consumption associated with wastewater treatment [35]. Additionally, biomass generation in microbial fuel cells is between 10% and 50% less than that produced by reducing the environmental effect and disposal costs of conventional wastewater treatment. Electrochemically active bacteria break down biodegradable organic compounds before releasing electrons to an external electron acceptor. Bacteria also carry out direct electron transfer, as well as shuttles and electrically conductive nanowires [36]. The study of bacterial electron transfer may lead to better understanding of the biomaterials and metabolic pathways involved, as well as improved microbial fuel cell output power. Electrochemistry, microbiology, material science, surface chemistry, and engineering disciplines such as reactor design, operation, and modeling are required to understand biofuel cell systems. Research collaboration and system integration may improve bioelectrochemical sensing systems for sustainable energy.

3.3 Volatile organic gases from agricultural processes monitoring

VOCs are a critical component in methanogenesis metabolic pathways, and their existence may signal the process's stability. Bioelectrochemical sensors for environmental or bioprocess monitoring have gained prominence in recent years. Despite the improvements, many challenges remain for VOC detection, including broad electrode spacing, low stability, biofouling at the cathode, and a narrow detection range. Zhang et al. created a new biosensor based on a three-chamber microbial electrochemical system for the purpose of monitoring acetate production throughout the anaerobic digestion process. Acetate was transported from the sample chamber to the anode through the anion-exchange membrane and subsequently oxidized as a substrate for current generation by the anodic biofilm driven by the concentration difference. This design eliminates the impact of fluctuating waste characteristics on the cathodic reaction. The amounts of acetate were changed to determine their effect on current density. It is considered the selectivity, the effect of sample temperature on the sample, and the external resistance. A connection between current densities and acetate concentrations (up to 160 mM) was found ($R^2 > 0.99$) for a certain reaction period (from 2 to 5 h). After 5 h of response, current density rose by about 20% when the sample temperature was elevated (e.g., 37°C and 55°C). As external resistance reduced, the detection range increased. The biosensor identified acetate concentrations in anaerobic digestion effluents that were within 24.2% of those reported by gas chromatography. Despite this, it may be advantageous to use the biosensor to detect acetate in environmental samples [37].

3.4 Diagnosis clinical

Despite significant advances in bioelectrochemical sensor research in recent years, their use in clinical diagnosis is still uncommon, with the exception of glucose biosensors, which account for more than 90% of the global biosensor market. The most serious issue is interfering molecules during measurements on real samples. Due to the fact that the majority of treatments are based on a limited number of clinical indications, high selectivity and accuracy are needed. The growing significance of tiny portable analytical devices for fast clinical testing, along with advancements in biosensor manufacturing, provides reason to be optimistic about biosensor commercialization success. The majority of biosensors on exhibit are amperometric in nature, indicating a trend in biosensor development. The most important job for developers is to ensure product stability, to eliminate unwanted interferences, and to economically viable their goods. Bioelectrochemical sensors have aided in the advancement of "clinical diagnosis for a variety of tests, including glucose, lactate, cholesterol, urea, creatinine, DNA, antigens, antibodies, and cancer indicators. The biosensors chosen were capable of operating on real samples and had high sensitivity, selectivity, and sensitivity. They are often intended to prevent interacting with undesirable components in the monitored systems" [8].

3.5 Monitoring the water's quality

The long-term provision of high-quality, safe water is a critical problem facing the world in the 21st century [38]. Numerous chemical and biological monitoring analyses are used to ensure that high-quality water is provided. However, "these techniques continue to have major drawbacks, including high prices, complex design, and physical and online limitations. The use of microbial fuel cell-based biosensors, a relatively new technology, has significant promise for monitoring the quality of water sources in real time [39]. Microbial fuel cells provide advantages in terms of ease of fabrication and efficiency for onsite sensing. Although some applications of microbial fuel cells for sensing, such as the biochemical oxygen demand sensor, have been studied previously, numerous research groups worldwide have recently presented novel practical applications of this technique that combine multidisciplinary scientific knowledge from the fields of materials science, microbiology, and electrochemistry" [40]. The most recent study is on the application of microbial fuel cells as possible biosensors for water quality monitoring, as well as the discovery of a spectrum of potentially hazardous analytes [41].

4 Conclusion

As described in this chapter, bioelectrochemical sensing has advanced applications and significant promise for sustainable energy and chemical synthesis. Bioelectrochemical sensing, along with green chemistry and environmental sustainability principles, has opened new vistas in sustainable waste disposal and environmental research. Testing bioelectrochemical sensors onsite is required to advance these technologies and make them commercially accessible. Another major issue is the creation of a simpler membraneless design. The impact of nanomaterials on the second and third generations of bioelectrochemical sensing is the promising research field's future. Since the discovery of electrotrophs—microorganisms that can directly receive electrons from electrodes for the reduction of terminal electron acceptors—a wide array of possible uses has been explored. Commercial microbial biosensors are the key to many new technologies, yet they are just the tip of the iceberg in terms of current academic research. This chapter may be used to develop next-generation bioelectrochemical sensors while also highlighting current achievements in the area.

References

[1] S. Chatterjee, N.P. Shetti, K. Raghava Reddy, 19—Fundamentals, recent advances, and perspectives of electrode materials for bioelectrochemical sensing applications, in: S. Kanchi, D. Sharma (Eds.), Nanomaterials in Diagnostic Tools and Devices, Elsevier, 2020, pp. 557–589, https://doi.org/10.1016/B978-0-12-817923-9.00019-5.

[2] D. Grieshaber, R. MacKenzie, J. Vörös, E. Reimhult, Electrochemical biosensors—sensor principles and architectures, Sensors 8 (3) (2008) 1400–1458. Accessed: Oct. 15, 2021. [Online]. Available https://www.ncbi.nlm.nih.gov/pmc/articles/PMC3663003/.

[3] S. Qazi, K. Raza, Chapter 4—Smart biosensors for an efficient point of care (PoC) health management, in: J. Chaki, N. Dey, D. De (Eds.), Smart Biosensors in Medical Care, Academic Press, 2020, pp. 65–85, https://doi.org/10.1016/B978-0-12-820781-9.00004-8.

[4] V. Naresh, N. Lee, A review on biosensors and recent development of nanostructured materials-enabled biosensors, Sensors 21 (2021), https://doi.org/10.3390/s21041109. no. 4, Art. no. 4.

[5] T. Lakshmipriya, S.C.B. Gopinath, 1—An introduction to biosensors and biomolecules, in: S.C.B. Gopinath, T. Lakshmipriya (Eds.), Nanobiosensors for Biomolecular Targeting, Elsevier, 2019, pp. 1–21, https://doi.org/10.1016/B978-0-12-813900-4.00001-4.

[6] J.P. Pender, et al., Electrode degradation in lithium-ion batteries, ACS Nano 14 (2) (2020) 1243–1295, https://doi.org/10.1021/acsnano.9b04365.

[7] R. Rajaram, J. Mathiyarasu, Chapter 1 The design and fabrication of disposable sensors: an overview, 2021, pp. 1–26, https://doi.org/10.1039/9781839163364-00001.

[8] Application of Electrochemical Biosensors in Clinical Diagnosis, doi: https://doi.org/10.1002/jcla.20500.

[9] J. Park, K.T. Kim, W.H. Lee, Recent advances in information and communications technology (ICT) and sensor technology for monitoring water quality, Water 12 (2020), https://doi.org/10.3390/w12020510. no. 2, Art. no. 2.

[10] Z. Panahi, L. Custer, J.M. Halpern, Recent advances in non-enzymatic electrochemical detection of hydrophobic metabolites in biofluids, Sens. Actuators Rep. 3 (2021) 100051, https://doi.org/10.1016/j.snr.2021.100051.

[11] N. Saifuddin, A.Z. Raziah, A.R. Junizah, Carbon nanotubes: a review on structure and their interaction with proteins, J. Chem. 2013 (2012) e676815, https://doi.org/10.1155/2013/676815.

[12] T. Zheng, et al., Progress and prospects of bioelectrochemical systems: electron transfer and its applications in the microbial metabolism, Front. Bioeng. Biotechnol. 8 (2020) 10, https://doi.org/10.3389/fbioe.2020.00010.

[13] F. Zhu, W. Wang, X. Zhang, G. Tao, Electricity generation in a membrane-less microbial fuel cell with down-flow feeding onto the cathode, Bioresour. Technol. 102 (15) (2011) 7324–7328, https://doi.org/10.1016/j.biortech.2011.04.062.

[14] N. Yasri, E.P.L. Roberts, S. Gunasekaran, The electrochemical perspective of bioelectrocatalytic activities in microbial electrolysis and microbial fuel cells, Energy Rep. 5 (2019) 1116–1136, https://doi.org/10.1016/j.egyr.2019.08.007.

[15] H. Chen, et al., Fundamentals, applications, and future directions of Bioelectrocatalysis, Chem. Rev. 120 (23) (2020) 12903–12993, https://doi.org/10.1021/acs.chemrev.0c00472.

[16] D. Ucar, Y. Zhang, I. Angelidaki, An overview of Electron acceptors in microbial fuel cells, Front. Microbiol. 8 (2017) 643, https://doi.org/10.3389/fmicb.2017.00643.

[17] V. Chaturvedi, P. Verma, Microbial fuel cell: a green approach for the utilization of waste for the generation of bioelectricity, Bioresour. Bioprocess. 3 (1) (2016) 38, https://doi.org/10.1186/s40643-016-0116-6.
[18] J. Blumberger, Recent advances in the theory and molecular simulation of biological Electron transfer reactions, Chem. Rev. 115 (20) (2015) 11191–11238, https://doi.org/10.1021/acs.chemrev.5b00298.
[19] F. Kracke, I. Vassilev, J.O. Krömer, Microbial electron transport and energy conservation – the foundation for optimizing bioelectrochemical systems, Front. Microbiol. 6 (2015) 575, https://doi.org/10.3389/fmicb.2015.00575.
[20] P. Choudhury, U.S. Prasad Uday, T.K. Bandyopadhyay, R.N. Ray, B. Bhunia, Performance improvement of microbial fuel cell (MFC) using suitable electrode and bioengineered organisms: a review, Bioengineered 8 (5) (2017) 471–487, https://doi.org/10.1080/21655979.2016.1267883.
[21] S.G.A. Flimban, I.M.I. Ismail, T. Kim, S.-E. Oh, Overview of recent advancements in the microbial fuel cell from fundamentals to applications: design, major elements, and scalability, Energies 12 (2019), https://doi.org/10.3390/en12173390. no. 17, Art. no. 17.
[22] M. Ramya, P. Senthil Kumar, A review on recent advancements in bioenergy production using microbial fuel cells, Chemosphere (2021) 132512, https://doi.org/10.1016/j.chemosphere.2021.132512.
[23] L. Ezziat, A. Elabed, S. Ibnsouda, S. El Abed, Challenges of microbial fuel cell architecture on heavy metal recovery and removal from wastewater, Front. Energy Res. 7 (2019) 1, https://doi.org/10.3389/fenrg.2019.00001.
[24] B.K. Behera, A. Varma, Gasoline-like biofuel, in: B.K. Behera, A. Varma (Eds.), Bioenergy for Sustainability and Security, Springer International Publishing, Cham, 2019, pp. 79–158, https://doi.org/10.1007/978-3-319-96538-3_2.
[25] B. Xu, Z. Ge, Z. He, Sediment microbial fuel cells for wastewater treatment: challenges and opportunities, Environ. Sci.: Water Res. Technol. 1 (3) (2015) 279–284, https://doi.org/10.1039/C5EW00020C.
[26] M.J. Angelaalincy, R. Navanietha Krishnaraj, G. Shakambari, B. Ashokkumar, S. Kathiresan, P. Varalakshmi, Biofilm engineering approaches for improving the performance of microbial fuel cells and bioelectrochemical systems, Front. Energy Res. 6 (2018) 63, https://doi.org/10.3389/fenrg.2018.00063.
[27] Y. Gao, J. An, H. Ryu, H.-S. Lee, Microbial fuel cells as discontinuous portable power sources: syntropic interactions with anode-respiring bacteria, ChemSusChem 7 (4) (2014) 1026–1029, https://doi.org/10.1002/cssc.201301085.
[28] G. Puthilibai, K.S. Gopika, R. Kaleeswari, Urinal microbial fuel cell: electricity generation—an eco friendly approach, in: 2020 International Conference on Power, Energy, Control and Transmission Systems (ICPECTS), 2020, pp. 1–6, https://doi.org/10.1109/ICPECTS49113.2020.9337028.
[29] D.R. Lovley, et al., Geobacter: The microbe electric's physiology, ecology, and practical applications, in: R.K. Poole (Ed.), Advances in Microbial Physiology, Vol. 59, Academic Press, 2011, pp. 1–100, https://doi.org/10.1016/B978-0-12-387661-4.00004-5.
[30] P. Arora, A. Sindhu, N. Dilbaghi, A. Chaudhury, Biosensors as innovative tools for the detection of food borne pathogens, Biosens. Bioelectron. 28 (1) (2011) 1–12, https://doi.org/10.1016/j.bios.2011.06.002.
[31] S.K. Malyan, S.S. Kumar, R.K. Fagodiya, P. Ghosh, A. Kumar, R. Singh, L. Singh, Biochar for environmental sustainability in the energy-water-agroecosystem nexus, Renew. Sust. Energ. Rev. (2021) 111379, https://doi.org/10.1016/j.rser.2021.111379.
[32] S.S. Kumar, S.K. Malyan, A. Kumar, S. Basu, N.R. Bishnoi, Microbial fuel cells technology: food to energy conversion by anode respiring bacteria, in: Environmental Concerns of 21st Century: Indian and Global Context, 2016, pp. 139–154.
[33] J.N. Edokpayi, J.O. Odiyo, O.S. Durowoju, Impact of wastewater on surface water quality in developing countries: a case study of South Africa, IntechOpen (2017), https://doi.org/10.5772/66561.
[34] M. Shemfe, K.S. Ng, J. Sadhukhan, Bioelectrochemical Systems for Biofuel (Electricity, Hydrogen, and Methane) and Valuable Chemical Production, 2017, https://doi.org/10.1201/b22351-11.

[35] C. Munoz-Cupa, Y. Hu, C. Xu, A. Bassi, An overview of microbial fuel cell usage in wastewater treatment, resource recovery and energy production, Sci. Total Environ. 754 (2021) 142429, https://doi.org/10.1016/j.scitotenv.2020.142429.
[36] S. Khandaker, S. Das, M. Tofazzal Hossain, A. Islam, M. Raza Miah, M. Rabiul Awual, Sustainable approach for wastewater treatment using microbial fuel cells and green energy generation—a comprehensive review, J. Mol. Liq. (2021) 117795, https://doi.org/10.1016/j.molliq.2021.117795.
[37] H. Sun, I. Angelidaki, S. Wu, R. Dong, Y. Zhang, The potential of bioelectrochemical sensor for monitoring of acetate during anaerobic digestion: focusing on novel reactor design, Front. Microbiol. 9 (2019) 3357, https://doi.org/10.3389/fmicb.2018.03357.
[38] S.K. Malyan, R. Singh, M. Rawat, M. Kumar, A. Pugazhendhi, A. Kumar, V. Kumar, S.S. Kumar, An overview of carcinogenic pollutants in groundwater of India, in: Biocatalysis and Agricultural Biotechnology, 2019, p. 101288, https://doi.org/10.1016/j.bcab.2019.101288.
[39] S.S. Kumar, V. Kumar, R. Kumar, S.K. Malyan, A. Pugazhendhi, Microbial fuel cells as a sustainable platform technology for bioenergy, biosensing, environmental monitoring, and other low power device applications, Fuel (2019) 115682, https://doi.org/10.1016/j.fuel.2019.115682.
[40] J.M. Sonawane, C.I. Ezugwu, P.C. Ghosh, Microbial fuel cell-based biological oxygen demand sensors for monitoring wastewater: state-of-the-art and practical applications, ACS Sens. 5 (8) (2020) 2297–2316, https://doi.org/10.1021/acssensors.0c01299.
[41] J.R. Kim, Application of bioelectrochemical process (BES) for electricity generation and sustainable wastewater Treatment, in: EKC 2009 Proceedings of the EU-Korea Conference on Science and Technology, Heidelberg, Berlin, 2010, pp. 17–23, https://doi.org/10.1007/978-3-642-13624-5_2.

CHAPTER 8

Microbial fuel cells as sustainable method of wastewater treatment

Bindu Mangla[a], Noor[a], Smita S. Kumar[b,c], Sudheesh K. Shukla[d,e], and Suresh Kumar[f]
[a]Department of Chemistry, J.C. Bose University of Science and Technology YMCA, Faridabad, India
[b]Department of Environmental Sciences, J.C. Bose University of Science & Technology, YMCA, Faridabad, Haryana, India
[c]Department of Environmental Studies, Central University of Haryana, Jant-Pali, Mahendergarh, Haryana, India
[d]School of Biomedical Engineering, Shobhit Institute of Engineering and Technology, Meerut, India
[e]Department of Chemical Sciences, University of Johannesburg, Doornfontein Campus, Johannesburg, South Africa
[f]Department of Chemistry, Kurukshetra University, Kurukshetra, India

Abbreviations

ATP	adenosine triphosphate
BOD	biological oxygen demand
COD	chemical oxygen demand
CW	cheese whey
DCMFC	double chambered microbial fuel cell
DET	direct electron transfer
EET	extracellular electron transfer
FPMFC	flat plate microbial fuel cell
LSCV	low scan rate cyclic voltammetry
MET	mediated electron transfer
MFC	microbial fuel cell
MSW	municipal solid waste
NAD^+	nicotinamide adenine dinucleotide
NADH	nicotinamide adenine dinucleotide dehydrogenase
PEM	proton exchange membrane
PMFC	plant MFC
SCMFC	single chambered microbial fuel cell
SMFC	sediment microbial fuel cell
TS	total solid
WWT	wastewater treatment

1 Microbial fuel cell (MFC)

A microbial fuel cell (MFC) is a bioelectrochemical device that is used for converting chemical energy contained by organic or inorganic compounds into electric current with the aid of microorganisms acting as a biocatalyst. MFC is one of the best-known and most appealing technologies for sustainable energy production and wastewater treatment

Algae Based Bioelectrochemical Systems for Carbon Sequestration, Carbon Storage, Bioremediation and Bioproduct Generation
https://doi.org/10.1016/B978-0-323-91023-1.00013-2

(WWT) [1–3]. In a basic layout, an MFC consists of two electrodes, cathode and anode, and an ion exchange membrane between them, generally a proton exchange membrane (PEM) [4,5]. For the generation of electric current, a variety of materials can be used, such as organic complex waste and natural organic matter [6–8]. The chambers containing the electrodes may be made from materials such as polycarbonate, glass or acrylic glass, plastic bottles or Plexiglas, or cans [9–11]. In MFCs, microorganisms oxidize organic substrates or organic pollutants into carbon dioxide, electrons, and protons. The generated carbon dioxide is released into the atmosphere; whereas electrons released from microbial oxidation of the substrate initially transfer to the anode surface through redox active proteins or cytochromes, and afterwards are transferred to the cathode through an external circuit. The generated protons are transported from anode to cathode through the PEM. Transported electrons get reduced in the cathodic chamber, where they combine with diffused protons through the PEM and oxygen, obtained from air, which produces water [12–18] (Fig. 1A).

Reaction at anode chamber:

$$CH_3COO^- + 2H_2O \rightarrow 2CO_2 + 7H^+ + 8e^- \ (E^{o\prime} = -0.29\,V)$$

Reaction at cathode chamber:

$$O_2 + 4H^+ + 4e^- \rightarrow 2H_2O \ (E^{o\prime} = -0.82\,V)$$

MFCs have advantages over other energy generation technologies, especially those based on bioenergy production, since it utilizes agriculture and industrial waste/residue and wastewater. A few of the advantages are compiled in Fig. 1B.

1.1 Materials

1.1.1 Anode

An anodic material should be conductive, chemically inert, and environmentally friendly. Carbon materials such as graphite plate, graphite rod, carbon cloth, carbon brush, graphite granules, graphite felt, and stainless steel, etc., should preferably be used for manufacturing of such electrodes, which are cheap and have large surface area. Bar electrodes adapted with conductive nanomaterials possess increased surface area and give enhance MFC performance [19–21] (Fig. 2).

1.1.2 Cathode

Oxygen, being cheap, having high redox potential, and being abundantly available, represents one of the best alternatives as an electron acceptor. When oxygen is used as an electron acceptor, a catalyst such as Pt is used on the cathode, which increases the rate of reduction. Ferricyanite can also be used, in which case no catalyst is needed.

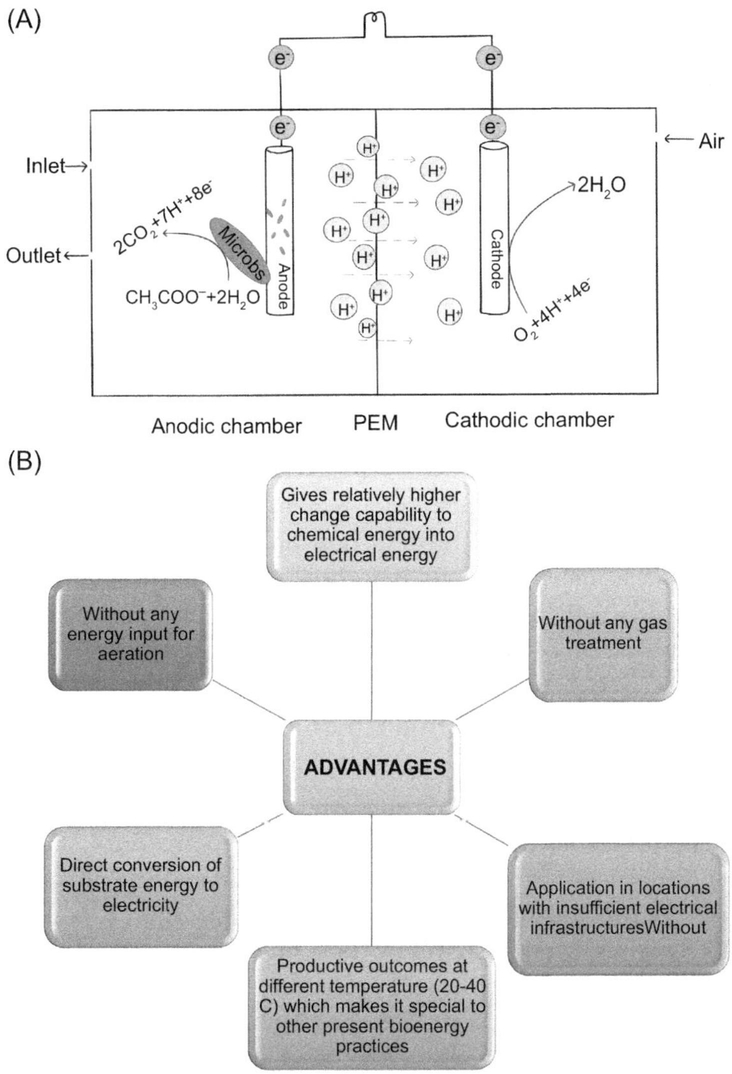

Fig. 1 (A) Basic structure of two chambered microbial fuel cell. (B) Advantages of MFC over other technologies.

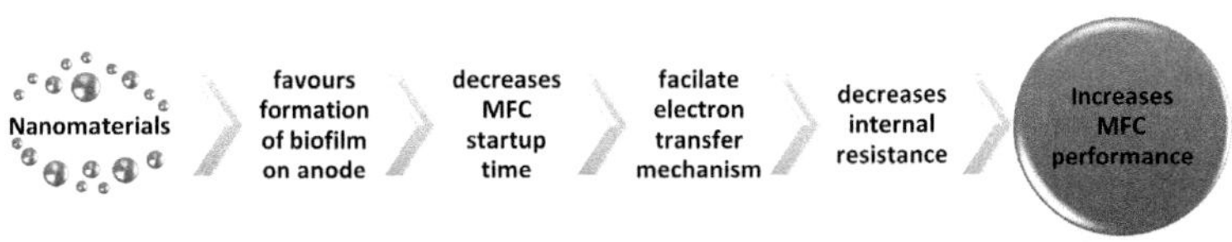

Fig. 2 Effect of nanoparticles on MFC performance.

Sometimes oxygen may diffuse from the cathodic to the anodic chamber, which decreases the performance of the MFC [22–26].

1.1.3 Membrane

The two chambers, cathode and anode, are separated by an ion exchange membrane, generally a proton exchange membrane (PEM), which allows protons to move through from anode to cathode. Nafion is considered to be a superior PEM material. UltraCMI-7000 is considered to be superior to Nafion, but is less affordable.

1.2 Types of MFC

The research and development of various types of MFCs has sparked interest because of its potential in ecologically friendly bioenergy generation from decomposable waste products. Design of MFCs may vary from single chambered to multichambered, from a two electrode system to a multielectrode system, etc. There are various types of MFC, including plant MFC, flat-plate MFC, horizontal MFC stack, vertical MFC stack, bipolar electrode stack, miniature microbial fuel cell, and sediment MFC [27–30], but the most commonly used MFCs are single chambered MFC, double chambered MFC, upflow MFC, and stacked MFC (Fig. 3).

1.3 Factors affecting the performance of MFCs

There are various factors that affect the performance of MFCs, such as electrode material, operating conditions, membrane type, substrate type, electron transfer mechanisms, design of system or type of MFC, microbial community, etc. (Fig.4). All these factors should be considered in order to design a productive apparatus that will achieve the maximum efficiency of MFC in WWT. However, high construction cost and slow rate of microbial activity limits these systems in working on a large scale [31]. Mostly, PEM is used in between two electrodes, which is a porous membrane with charged side walls to support the movement of protons on the way to the cathode. Nafion is the most commonly used PEM due to its proton-selective permeable nature. It has been found that the membrane surface area to volume of MFC ratio is important for the electric current of the system: the greater the surface area of the electrode, the higher the performance.

Modification with nanoparticles in the anode leads to increase in power output, as it supports the formation of biofilm and enhances the electron transfer rate, which in turn increases the output [32] (Fig. 5C). The type of substrate and its concentration affects the output of MFC. There are two ways to categorize a substrate: first, according to its degradability, and second, according to its simplicity or complexity [33,34] (Fig. 5A). Microbes present on the anode surface oxidize organic substrates to finally yield electricity; hence the choice of microbial community is vital for the performance of MFCs. Consequently, properties such as the nature and stability of microbes, the extracellular-electron

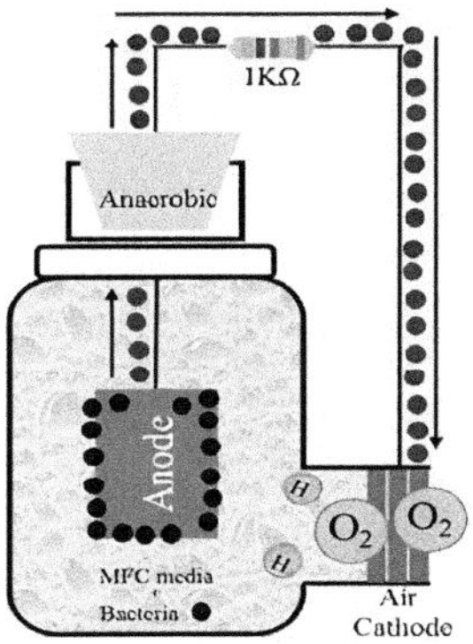

SINGLE CHAMBER MFC
- One chamber holding both cathode and anode.
- Anode kept far from PEM and cathode near the PEM.
- Easy ,cheap and generates more power as compare to double chamber.
- Requires low set-up cost.

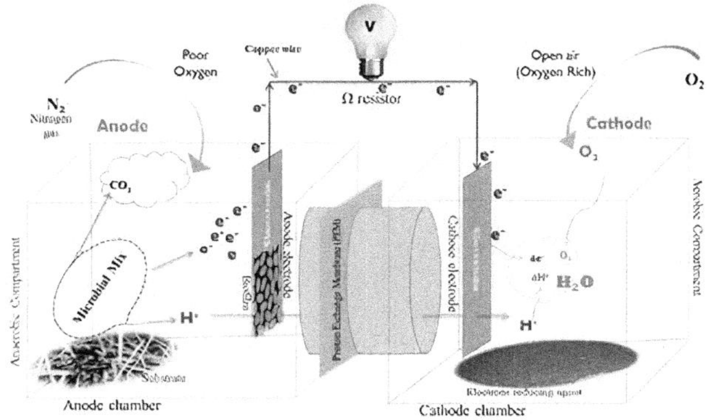

DOUBLE CHAMBER MFC
- Simplest design.
- Contains two chamber cathode and anode and separated by PEM.
- Cube or bottle shape
- Works in batch mode.
- Both chambers can be maintained with different operating states.
- Total output can be enhances by varying factors.

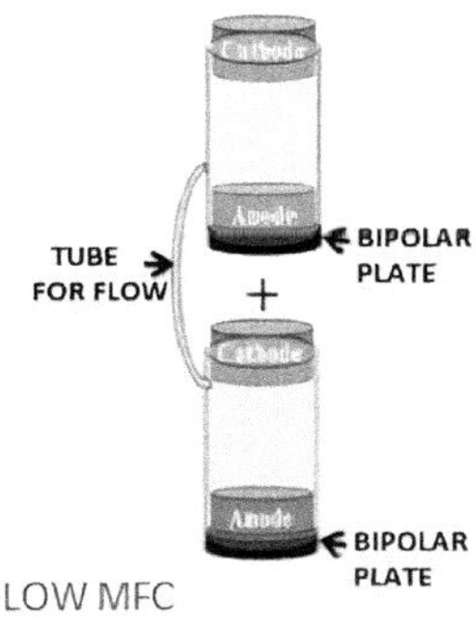

UPFLOW MFC
- Cylindrical shape
- Cathode chamber at top and anode chamber at bottom.
- Anode contains substrate that moves to cathode.
- A gradient is generated between both electrodes which is beneficial for performance of MFC.
- Glass bead layer and glass wool is shared to both the chambers.
- Easily extended and works in continuous operation.

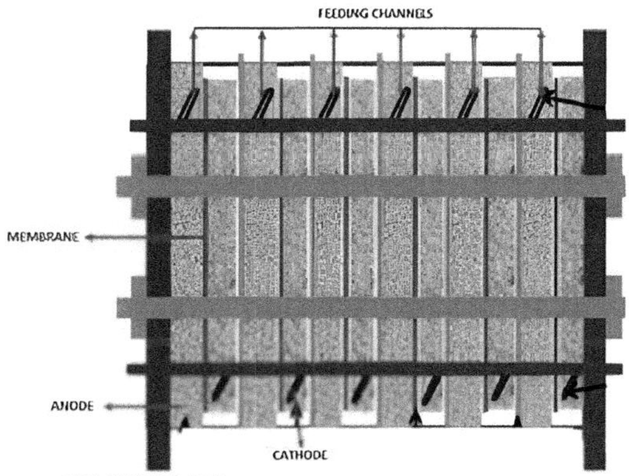

STACKED MFC
- Some MFC are combined in series or in parallel to increase the power output by multiplying individual current output.
- Due to loss of some voltage the output voltage is less then the total voltages of individual cell.
- Proven that parallel connection gives more output current as compare to the series.
- Connection and the parallel connection maximize COD removal and WWT efficiency.

Fig. 3 Types of MFC with their key features.

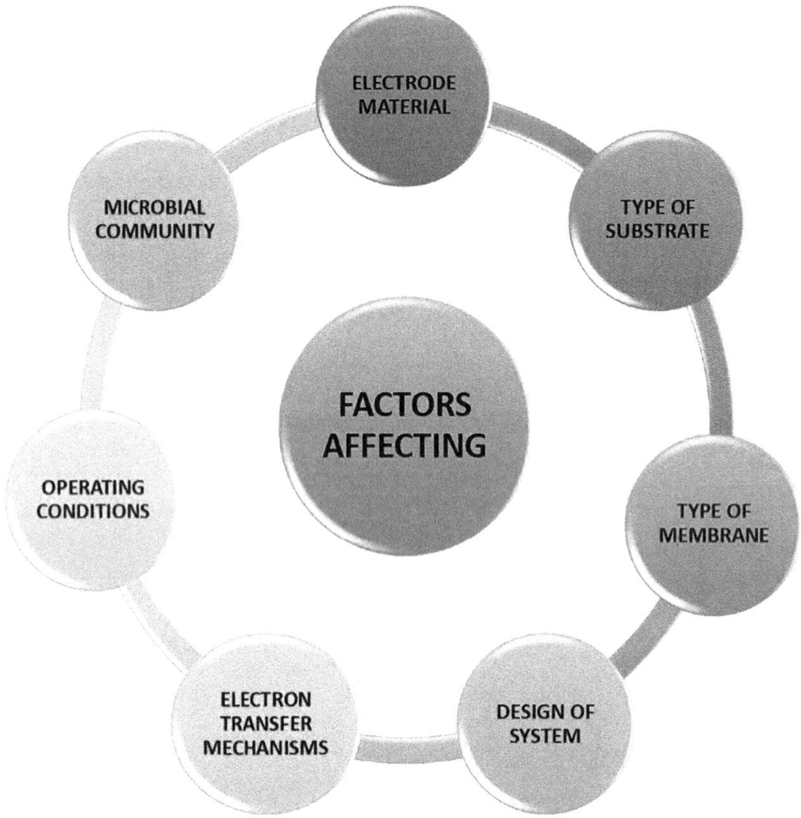

Fig. 4 Factors affecting functioning of MFC.

transfer towards the anode that initiates the reaction, and the activity of microbes have been researched. Organic matter is oxidized by microorganisms in the anodic chamber, and electrons are created by microbial metabolism. Adenosine triphosphate (ATP), also known as the biological energy unit of the microbial cell, is produced by repeated oxidation and reduction processes in all types of MFC like, bacterial fuel cells (microbe used: bacteria), fungal fuel cells (microbe used: fungi), and algal fuel cells (microbe used: algae) [35].

Long-term MFC operation causes an increase in anolyte proton concentration, which restricts or delays the proton flow from PEM. Due to the increase in the proton concentration, the anodic chamber becomes more acidic, which produces unfavorable conditions for microbial growth and, on other side, due to lack of protons, the cathodic chamber becomes alkaline [36,37]. A pH range of 6–9 is found to be suitable for proper microbial growth and high output current; at high pH, oxygen reduction potential decreases and decreases the output current [38,39], whereas at low pH, microbial activity

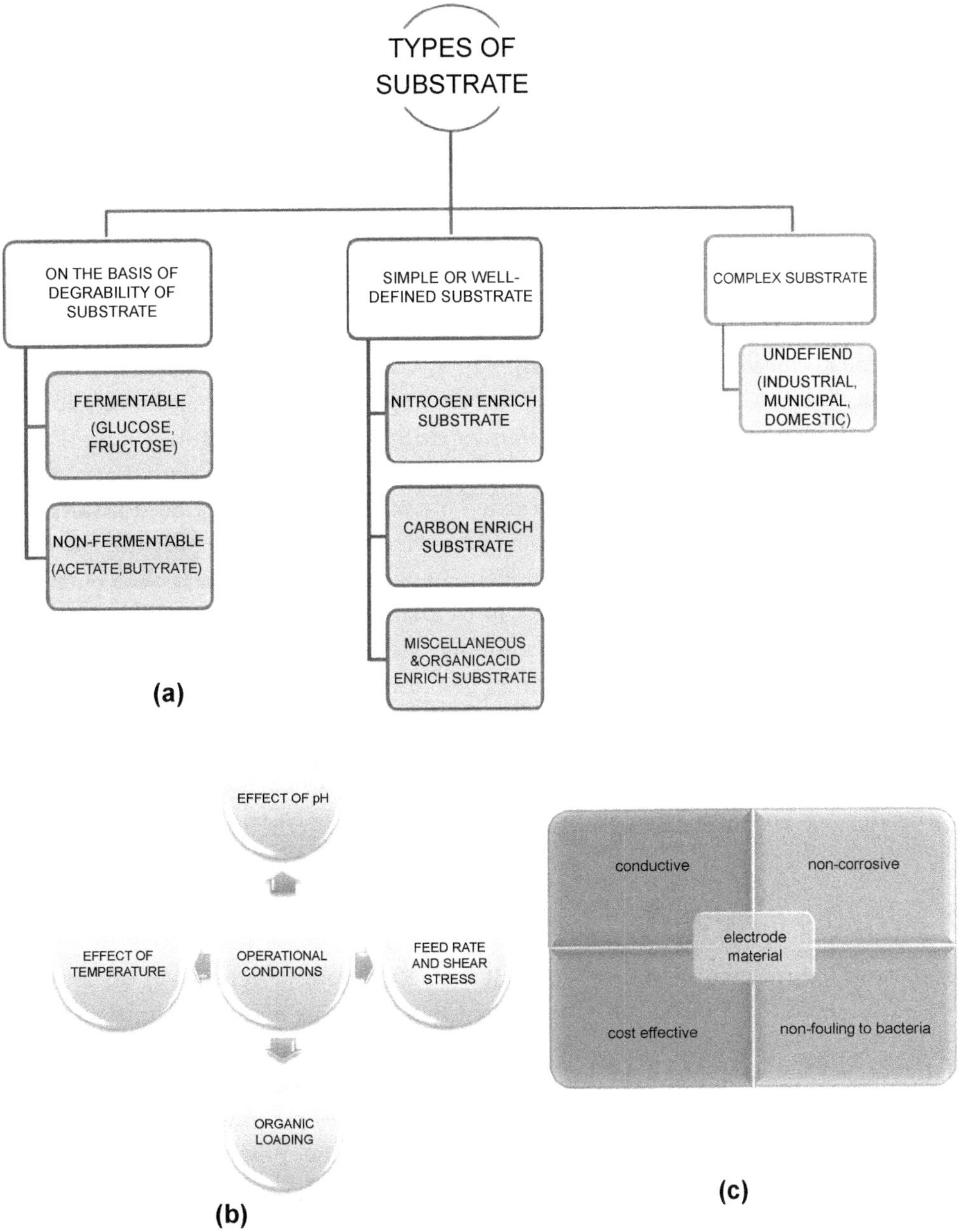

Fig. 5 MFC performance affecting factors: (A) types of substrate, (B) operational conditions, (C) electrode materials.

decreases in anolyte, which affects the formation of biofilm and output current. Temperature is a crucial parameter, as it affects the kinetics and thermodynamics of mass transfer and microbial metabolism of the MFC. Temperature also affects the start-up time of MFC: as the temperature increases, start-up time decreases, which leads to the formation of biofilm, which ultimately impacts the output current generation [40]. Flow rate has also been found to have an effect on MFC function: high flow rates bring down not only the output current but also columbic efficiency and COD elimination [41,42]. Production of thicker biofilms is observed with low shear rate. Distinct microbial communities can develop at the anode when MFC is operated at distinct shear rates. Higher shear rates in MFC have been found to reduce microbial variety, resulting in the production of homogeneous biofilms [43,44] (Fig. 5B).

Extracellular electron transfer takes place between the microbial community and the electrodes, either as mediated electron transfer (MET) or direct electron transfer (DET). In a DET mechanism, the electrode remains in direct physical contact with the microbial cell membrane and electron transfer is carried out by microbes via the outside surface of proteins, known as outer membrane cytochromes or conductive pili. C-type cytochromes play an important part in the movement of electrons from iner cell to outer cell, and their advantages are evident. DET can be characterized by techniques like LSCV, Raman microscopy, environmental-scanning electron microscopy, and electron-impedance spectroscopy. MET, sometimes described as an indirect method of electron transfer, needs a transporter or redox carrier, which facilitates the transfers of electron to the terminal acceptor. A good mediator is generally easily accessible and inexpensive, nontoxic to microbes, able to easily move through microbes' outer surfaces, and will create a sufficient amount of redox potential, which works on the electron movement.

2 Electricity generation

In most of the research work done on sustainable methods of generation of electricity [45], a substrate that can be entirely oxidized into electrons to attain output current is considered to be significant [46]. A great deal of effort has been put into enhancing the output current; for example, modifications in MFC system to decrease the internal resistance, the use of inexpensive PEM, low cost cathode and anode with large surface area, and even alterations made with nanomaterials and other physical or chemical treatment techniques [47].

3 Wastewater treatment

A microbial fuel cell utilizes microbes as biocatalyst to oxidize organic matter and, selectively, inorganic matter in an anodic chamber and generates energy in the form of electricity. Bioenergy production with MFC provides the dual benefit of power generation using wastewater as a substrate material, so eradicating the problem of environmental pollution to some extent. MFC can be used to treat wastewater from a variety of sources, including domestic, industrial, agricultural, and sewage, although it is impossible to clean the toxic wastewater completely but remove COD to a level to meet the discharge criteria into natural surroundings [48,49]. Wastewater obtained from different sources/processes will have a different composition and, therefore, will require particular kinds of microbes for its digestion and produce distinct power density. Some commonly used wastewaters for MFC are discussed in the present study.

4 Use of different wastewaters for MFC

MFC technology has improved dramatically in terms of treatment efficiency and power density over the last few years. MFCs have demonstrated outstanding success in the removal of contaminants from wastewater, for example, through chemical oxygen demand (COD) [50]. The substrate is thought to be the most important biological element in MFC, as energy production is heavily reliant on it [51]. Thermophilic metabolism and high-temperature cellulose biodegradability might be employed to favor MFC in order to reach larger power densities. Different types of wastewater, for example, domestic wastewater and industrial wastewater, such as starch processing wastewater, food processing wastewater, chocolate factory wastewater, etc., have been employed to generate energy using MFC [52,53].

4.1 Agricultural and agro-industrial waste

Agro-processing industrial waste and agricultural residue is nutrient-rich, with carbon-based compounds that are easily biodegradable due to the presence of a large quantity of organic matter along with micro- and macronutrients that are ideal for microbial growth. Agricultural and agro-industrial waste can be further classified depending upon availability and source; commonly, it is grouped into four generations [54]:

- *First generation*:
  - Includes different types of food crops, e.g., sorghum, corn, and wheat
  - Straight implementation as primary feedstock
  - Produces energy and generates different products
  - Limitation: high fuel production investment required
- *Second generation*:
  - Includes lignocellulosic waste, such as wood chips, organic waste, crop residues, and sugarcane bagasse
  - Generates bioenergy
  - Superior to first generation biomass

- *Third generation*:
  - Includes microalgal biomass
  - Planting in open pounds or in lagoons utilizing nitrogenous compounds
- *Fourth generation*:
  - Algae is produced from emissions control techniques or cleaner disposal methods.
  - Generation of high value products
  - Larger polymeric hydrocarbon content

4.2 Domestic or municipal wastewater

A lot of feedstock and contaminants with low toxicity are generated from municipal and household wastewater, which can be utilized in power production in MFC [27,55,56]. The costs associated with managing sludge in sewage treatment facilities are often a significant expense [57]; Lowering the cost of sludge treatment can reduce the overall expenses of wastewater treatment (WWT). Additionally, wastewater treatment plants yield a high concentration of organic materials that can be effectively utilized in Microbial Fuel Cells (MFCs) [58].

4.3 Food and food-processing industry wastewater

A lot of food is wasted at the end of the food chain, which is rich in carbohydrates content. Municipal solid waste (MSW) consists of approx. 27% food waste. Wastewater produced by protein food industry is nontoxic with high BOD and organic content, owing to which, COD removal increases [59–62]. Brewery wastewater has high COD, and is nontoxic and rich in organics content, such as sugar, starch, and protein components that make it a good substrate for biocatalysis in MFC [63].

4.4 Dairy industry wastewater

Wastewater from the dairy industry is rich in organic complexes such as lipids, proteins, and polysaccharides, which, on hydrolysis, give fatty acids, organic acids, and sugar. A prime nutritionally rich product of the dairy industry is cheesy whey (CW). Using DCMFC, a power density of $1.3 \pm 0.5\,W/m^2$ and COD removal of $59.0\% \pm 9.3$ can be obtained. With other dairy wastes, various type of MFC can be used, which result in different power density generation and COD removal percentages. For example SCMFC provides power density and COD of $1.1\,W/m^2$ and 95.49%, respectively. In DCMFC power density and COD are $161\,mW/m^2$ and 90% respectively. In annular single chambered MFC, power density and COD are $20.2\,W/m^2$ and 91%, respectively [64–68].

4.5 Palm oil mill effluent

Palm oil is highly digestive and nutritive oil available in two types: solid palm kernels and crude palm oil. The palm oil agro-industry produces palm oil mill effluent through many extraction techniques. It has been found that anaerobic sludge with the aid of DCMFC gives a power density of 45–304 mW/m^2 and COD removal of 0.8%–45% [69–71].

4.6 Protein food industry wastewater

Wastewater generated from the protein food industry is nontoxic, with high levels of organic content compounds like sugar and starch. It also releases few hazardous products with high BOD content compounds. With DCMFC, power density and COD removal of 81 mW/m^2 and 95% respectively can be achieved [62,72].

4.7 Industrial pomace

Pomace is the dry or pulpy residue of material after extracting liquid juice/oil from it. Tomato pomace, grape pomace, olive pomace, etc., are rich in organic carbonaceous content, which is possibly suitable for the use as a feedstock in MFC to generate electricity. Power density of 132 mW/m^2 is produced from tomato industry pomace [73].

4.8 Brewery wastewater

Brewer's effluent is a high-strength wastewater that can be utilized to produce power. It is produced in enormous quantities during the brewing of beer, which involves multiple processing steps such as saccharification, fermentation, and other procedures [74]. SCMFC is used in brewery wastewater diluted with domestic wastewater that gives power density and COD of 30 mW/m^2 and 90.4%, respectively [75].

4.9 Steroidal drug production wastewater

The manufacture of hydrocortisone is a crucial step in the development of steroidal drugs. A substantial amount of highly toxic refining wastewater is created during the acid hydrolysis stage. It generates power density and COD of 22.3 W/m^3 and 82%, respectively [76–78].

4.10 Domestic and municipal wastewater

Energy production from domestic and municipal wastewater is important with reference to waste remediation and bioenergy production. SCMFC with domestic wastewater generates power density and COD removal of 26 mW/m^2 and 80%, respectively [20,27,79]. Air-cathode MFC produces power density and COD removal of 22.5 W/m^2 and 83%, respectively [80]. Power density was enhanced (124.6 mW/m^2) when compared to residential wastewater alone. In a dual chambered MFC, the residue from the incineration of municipal solid waste (MSW) was also examined as substrate, yielding a power density of 80 mW/m^2 [81].

4.11 Molasses wastewater

Molasses is obtained during refining of sugarcane in mills, and consists of sugar, vitamins, and minerals such as calcium. However, sugarcane factory molasses acts as a pollutant due to high COD content [82]. Single chambered cuboid MFC used for molasses wastewater can achieve 1410 mW/m^2 power density and 53.2% COD removal. SCMFC used in distillery wastewater (molasses based) gives power density and COD removal of 124.35 mW/m^2 and 72.84%, respectively. However molasses wastewater mixed with sewage on SCMFC can generate power density of 382 mW/m^2 and COD removal of 59% [83–85].

4.12 Rice straw

Lignocellulosic biomass, with varying ratios of organic carbon, with its residual components and their hydrolysates serve as substrates within Microbial Fuel Cells (MFCs) to generate power. Specifically, powdered rice straw is employed in an H-type MFC configuration, resulting in a power density of 145 mW/m^2 [86].

4.13 Yoghurt wastewater

For anaerobic sludge, yoghurt wastewater is utilized as fuel. When used in MFC, it gives a power density of 54 mW/m^2 and, when utilized as compost leachate, then it gives power density of 37 mW/m^2 [87].

4.14 Vegetable waste

Vegetable waste is a promising substrate, which can produce energy through various forms of MFC. The waste is generated from restaurants, markets, and vegetable packaging industries at the time of cutting and washing of vegetables [88]. The vegetable waste in SCMFC generates a power density of 57.38 mW/m^2 and COD is 62.86% [89]. SCMFC used with fermented vegetable waste provides power density and COD of 111.76 mW/m^2 and 80%, respectively [90].

4.15 Confectionary industry wastewater

Wastewater generated from chocolate/candy, or what is commonly called the confectionary industry, consists of high biological oxygen demand (BOD), COD, and total solid (TS) content. Chocolate industrial wastewater examined in DCMFC resulted in power density of 1500 mW/m^2 and COD removal of 74.77% [91].

5 Challenges of using MFC

The advancement of MFC technology over the past several years has significantly contributed to upholding the WWT and power generation in the forefront. Released organic waste that is harmful for both land-based and aquatic life can be used in MFC for bioenergy generation, and wastewater treatment can occur simultaneously and naturally. The MFC based on WWT has various challenges, including selection of substrate, and optimum temperature and pH for microorganisms. There are challenges to apply MFC on a large scale because of high capital cost and low energy output. The materials used in MFC as electrode materials, membrane, and required catalyst are of high cost [92].

6 Conclusion and future aspects

The gap between freshwater availability and wastewater generation is widening exponentially, thus creating a growing imbalance in ecology and the environment. Wastewater treatment using sustainable methods is the need of the day to meet emergent demand and to protect the depleting environment. MFC is one of the promising alternatives for wastewater treatment and electricity production in which organic/inorganic waste is used by microbes as feedstock in the anodic chamber of the MFC. MFCs work on the principle of bioelectrocatalysis, where microbes convert chemical energy stored in the feedstock to electrical energy. The interesting thing about MFC is that it is eco-friendly, as it can treat wastewater, remove pollutants, decrease COD, and provide a safe environment while achieving electricity generation.

The efficiency of MFC technology for wastewater treatment and power generation needs improvement to materialize it and scale up. More technical support is required to choose the best combination of MFC designs, electrodes, substrates, and biocatalysts for higher output. MFCs' interconnection capabilities with other similar-purpose technologies should be investigated in order to expand the chances of their assimilation for enhancing existing wastewater treatment facilities and so promote MFCs' practical application prospects [93]. Local governments and regulatory bodies should collaborate in investigating techniques to generate bioelectricity from farm waste using MFC.

References

[1] P. Aelterman, K. Rabaey, T.H. Pham, N. Boon, W. Verstraete, Continuous electricity generation at high voltages and currents using stacked microbial fuel cells, Environ. Sci. Technol. 40 (2006) 3388–3394, https://doi.org/10.1021/es0525511.

[2] S.S. Kumar, S.K. Malyan, A. Kumar, S. Basu, N.R. Bishnoi, Microbial fuel cells technology: food to energy conversion by anode respiring bacteria, in: Environmental Concerns of 21st Century: Indian and Global Context, 2016, pp. 139–154.

[3] R. Kumar, L. Singh, Z.A. Wahid, Exoelectrogens: recent advances in molecular drivers involved in extracellular electron transfer and strategies used to improve it for microbial fuel cell applications, Renew. Sust. Energ. Rev. 56 (2016) 1322–1336, https://doi.org/10.1016/j.rser.2015.12.029.

[4] S.S. Kumar, V. Kumar, R. Kumar, S.K. Malyan, N.R. Bishnoi, Ferrous sulfate as an in-situ anodic coagulant for enhanced bioelectricity generation and COD removal from landfill leachate, Energy 176 (2019) 570–581, https://doi.org/10.1016/j.energy.2019.04.014.

[5] M.J. Angelaalincy, R. Navanietha Krishnaraj, G. Shakambari, B. Ashokkumar, S. Kathiresan, P. Varalakshmi, Biofilm engineering approaches for improving the performance of microbial fuel cells and bioelectrochemical systems, Front. Energy Res. 6 (2018) 63, https://doi.org/10.3389/fenrg.2018.00063.
[6] S.K. Chaudhuri, D.R. Lovley, Electricity generation by direct oxidation of glucose in mediator less microbial fuel cells, Nat. Biotechnol. 21 (2003) 1229–1232, https://doi.org/10.1038/nbt867.
[7] K. Inoue, C. Leang, A.E. Franks, T.L. Woodard, K.P. Nevin, D.R. Lovley, Specific localization of the c-type cytochrome OmcZ at the anode surface in current-producing biofilms of Geobacter sulfurreducens, Environ. Microbiol. Rep. 3 (2010) 211–217, https://doi.org/10.1111/j.1758-2229.2010.00210.x.
[8] S.K. Malyan, S.S. Kumar, L. Singh, R. Singh, D.A. Jadhav, V. Kumar, Bioelectrochemical systems for removal and recovery of heavy metals, in: Bioremediation, Nutrients, and Other Valuable Product Recovery, Elsevier, 2021, pp. 185–203, https://doi.org/10.1016/B978-0-12-821729-0.00005-1.
[9] A. Parkash, S. Aziz, M. Abro, S.A. Soomro, A. Kousar, Design and fabrication of microbial fuel cell using cow manure for power genration, Sci. Int. 27 (2015) 4235–4238.
[10] Z. Du, H. Li, T. Gu, A state of the art review on microbial fuel cells: a promising technology for wastewater treatment and bioenergy, Biotechnol. Adv. 25 (5) (2007) 464–482, https://doi.org/10.1016/j.biotechadv.2007.05.004.
[11] L.A. Obasi, C.C. Opara, A. Oji, Performance of cassava starch as a proton exchange membrane in a dual chambered microbial fuel cell, Int. J. Eng. Sci. Technol. 4 (1) (2012).
[12] A.P. Borole, G. Reguera, B. Ringeisen, Z.W. Wang, Y. Feng, B.H. Kim, Electroactive biofilms: current status and future research needs, Energy Environ. Sci. 4 (2011) 4813–4834, https://doi.org/10.1039/C1EE02511B.
[13] R. Kumar, L. Singh, Z.A. Wahid, M. Fadhil, Exoelectrogens in microbial fuel cells towards bioelectricity generation: a review, Int. J. Energy Res. 39 (2015) 1048–1067, https://doi.org/10.1002/er.3305.
[14] K. Chandrasekhar, A. Kadier, G. Kumar, R.A. Nastro, V. Jeevitha, Challenges in microbial fuel cell and future scope, in: Microbial Fuel Cell, Springer, Cham, 2018, pp. 483–499, https://doi.org/10.1007/978-3-319-66793-5_25.
[15] D.R. Lovley, Bug juice: harvesting electricity with microorganisms, Nat. Rev. Microbiol. 4 (7) (2006) 497–508, https://doi.org/10.1038/nrmicro1442.
[16] Q. Zhang, J. Hu, D.J. Lee, Microbial fuel cells as pollutant treatment units: research updates, Bioresour. Technol. 217 (2016) 121–128. https://doi.org/10.1016/j.biortech.2016.02.006.
[17] K. Rabaey, W. Verstraete, Microbial fuel cells: novel biotechnology for energy generation, Trends Biotechnol. 23 (6) (2005) 291–298. https://doi.org/10.1016/j.tibtech.2005.04.008.
[18] H. Wang, J.D. Park, Z.J. Ren, Practical energy harvesting for microbial fuel cells: a review, Environ. Sci. Technol. 49 (6) (2015) 3267–3277, https://doi.org/10.1021/es5047765.
[19] B.E. Logan, J.M. Regan, Microbial fuel cells—challenges and applications, Environ. Sci. Technol. 40 (2006) 5172–5180, https://doi.org/10.1021/es0627592.
[20] H. Liu, R. Ramnarayanan, B.E. Logan, Production of electricity during wastewater treatment using a single chamber microbial fuel cell, Environ. Sci. Technol. 38 (2004) 2281–2285, https://doi.org/10.1021/es034923g.
[21] A. Bergel, D. Feron, A. Mollica, Catalysis of oxygen reduction in PEM fuel cell by seawater biofilm, Electrochem. Commun. 7 (2005) 900–904, https://doi.org/10.1016/j.elecom.2005.06.006.
[22] Y. Wang, A. Wang, A. Zhou, W. Liu, L. Huang, M. Xu, H. Tao, Electrode as sole electrons donor for enhancing decolorization of azo dye by an isolated WYZ-2, Bioresour. Technol. 152 (2014) 530–533, https://doi.org/10.1016/j.biortech.2013.11.001.
[23] S.S. Kumar, A. Rai, R. Singh, V. Kumar, D. Kumar, J. Kumar, A. Kumar, S.K. Malyan, Bioelectroremediation technologies in remediation of environmental pollutants: challenges and future prospects, in: Bioremediation for Environmental Sustainability, 2021, pp. 147–165, https://doi.org/10.1016/B978-0-12-820318-7.00007-1.

[24] D.R. Bond, D.R. Lovley, Electricity production by Geobacter sulfurreducens attached to electrodes, Appl. Environ. Microbiol. 69 (2003) 1548–1555, https://doi.org/10.1128/AEM.69.3.1548-1555.2003.
[25] A. Rhoads, H. Beyenal, Z. Lewandowski, Microbial fuel cell using anaerobic respiration as an anodic reaction and biomineralized manganese as a cathodic reactant, Environ. Sci. Technol. 39 (2005) 4666–4671, https://doi.org/10.1021/es048386r.
[26] L. Huang, X. Chai, X. Quan, B.E. Logan, G. Chen, Reductive dechlorination and mineralization of pentachlorophenol in biocathode microbial fuel cells, Bioresour. Technol. 111 (2012) 167–174, https://doi.org/10.1016/j.biortech.2012.01.171.
[27] B. Min, B.E. Logan, Continuous electricity generation from domestic wastewater and organic substrates in a flat plate microbial fuel cell, Environ. Sci. Technol. 38 (2004) 5809–5814.
[28] D. Strik, H.V.M. Hamelers, J. Snel, C.J. Buisman, Green electricity production with living plants and bacteria in a fuel cell, Int. J. Energy Res. 32 (2008) 870–876.
[29] C. Donovan, A. Dewan, H. Peng, D. Heo, H. Beyenal, Power management system for a 2.5W remote sensor powered by a sediment microbial fuel cell, J. Power Source 196 (3) (2011) 1171–1177.
[30] C.E. Reimers, L.M. Tender, S. Fertig, W. Wang, Harvesting energy from the marine sediment—water Interface, Environ. Sci. Technol. 35 (2001) 192–195.
[31] M. Angelaalincy, N. Senthilkumar, R. Karpagam, G.G. Kumar, B. Ashokkumar, P. Varalakshmi, Enhanced extracellular polysaccharide production and self-sustainable electricity generation for PAMFCs by Scenedesmus sp. SB1. ACS, Omega 2 (7) (2017) 3754–3765, https://doi.org/10.1021/acsomega.7b00326.
[32] F.A. Alatraktchi, Y. Zhang, I. Angelidaki, Nanomodification of the electrodes in microbial fuel cell: impact of nanoparticle density on electricity production and microbial community, Appl. Energy 116 (2014) 216–222, https://doi.org/10.1016/j.apenergy.2013.11.058.
[33] A. Parkash, Microbial fuel cells: a source of bioenergy, J. Microb. Biochem. Technol. 8 (3) (2016) 247–255, https://doi.org/10.4172/1948-5948.1000293.
[34] H.M. Singh, A.K. Pathak, K. Chopra, V.V. Tyagi, S. Anand, R. Kothari, Microbial fuel cells: a sustainable solution for bioelectricity generation and wastewater treatment, Biofuels 10 (1) (2019) 11–31, https://doi.org/10.1080/17597269.2017.1413860.
[35] S. Choi, Microscale microbial fuel cells: advances and challenges, Biosens. Bioelectron. 69 (2015) 8–25, https://doi.org/10.1016/j.bios.2015.02.021.
[36] S. Venkata Mohan, G. Velvizhi, J.A. Modestra, S. Srikanth, Microbial fuel cell: critical factors regulating bio-catalyzed electrochemical process and recent advancements, Renew. Sust. Energ. Rev. 40 (2014) 779–797, https://doi.org/10.1016/j.rser.2014.07.109.
[37] Z. He, N. Wagner, S.D. Minteer, L.T. Angenent, The upflow microbial fuel cell with an interior cathode: assessment of the internal resistance by impedance spectroscopy, Environ. Sci. Technol. 40 (2006) 5212–5217, https://doi.org/10.1021/es060394f.
[38] N.T. Phung, J. Lee, K.H. Kang, I.S. Chang, G.M. Gadd, B.H. Kim, Analysis of microbial diversity in oligotrophic microbial fuel cells using 16S rDNA sequences, FEMS Microbiol. Lett. 233 (2004) 77–82, https://doi.org/10.1016/j.femsle.2004.01.041.
[39] J. Niessen, U. Schröder, F. Scholz, Exploiting complex carbohydrates for microbial electricity generation-a bacterial fuel cell operating on starch, Electrochem. Commun. 6 (2004) 955–958, https://doi.org/10.1016/j.elecom.2004.07.010.
[40] Y. Yang, Y. Ding, Y. Hu, B. Cao, S.A. Rice, S. Kjelleberg, H. Song, Enhancing bidirectional electron transfer of Shewanella oneidensis by a synthetic flavin pathway, ACS Synth. Biol. 7 (2015) 815–823, https://doi.org/10.1021/sb500331x.
[41] B.E. Logan, C. Murano, K. Scott, N.D. Gray, I.M. Head, Electricity generation from cysteine in a microbial fuel cell, Water Res. 39 (2005) 942–952, https://doi.org/10.1016/j.watres.2004.11.019.
[42] M. Sun, D.D. Reible, G.V. Lowry, K.B. Gregory, Effect of applied voltage, initial concentration, and natural organic matter on sequential reduction/oxidation of nitrobenzene by graphite electrodes, Environ. Sci. Technol. 46 (2012) 6174–6181, https://doi.org/10.1021/es300048y.

[43] B.R. Ringeisen, E. Henderson, P.K. Wu, J. Pietron, R. Ray, B. Little, J.C. Biffinger, J.M. Jones-Meehan, High power density from a miniature microbial fuel cell using Shewanella oneidensis DSP10, Environ. Sci. Technol. 40 (2006) 2629–2634, https://doi.org/10.1021/es052254w.

[44] S.V. Raghavulu, J. Annie Modestra, K. Amulya, C.N. Reddy, S. Venkata Mohan, Relative effect of bioaugmentation with electrochemically active and nonactive bacteria on bioelectrogenesis in microbial fuel cell, Bioresour. Technol. 146 (2013) 696–703, https://doi.org/10.1016/j.biortech.2013.07.097.

[45] R. Orellana, J.J. Leavitt, L.R. Comolli, R. Csencsits, N. Janot, K.A. Flanagan, A.S. Gray, C. Leang, M. Izallalen, T. Mester, U (VI) reduction by a diversity of outer surface c-type cytochromes of Geobacter sulfurreducens, Appl. Environ. Microbiol. 79 (2013) 6369–6374, https://doi.org/10.1128/AEM.02551-13.

[46] G. Reguera, K.D. McCarthy, T. Mehta, J.S. Nicoll, M.T. Tuominen, D.R. Lovley, Extracellular electron transfer via microbial nanowires, Nature 435 (7045) (2005) 1098–1101, https://doi.org/10.1038/nature03661.

[47] M. Zhou, H. Wang, J.D. Hassett, T. Gu, Recent advances in microbial fuel cells (MFCs) and microbial electrolysis cells (MECs) for wastewater treatment, bioenergy and bioproducts, J. Chem. Technol. Biotechnol. 88 (2013) 508–518, https://doi.org/10.1002/jctb.4004.

[48] S. Oh, J. Kim, G. Premier, T. Lee, K. Changwon, W. Sloan, Sustainable wastewater treatment: how might microbial fuel cells contribute, Biotechnol. Adv. 28 (2010) 871–881, https://doi.org/10.1016/j.biotechadv.2010.07.008.

[49] M. Azam, S.S. Jahromy, W. Raza, N. Raza, S.S. Lee, K.-H. Kim, F. Winter, Status, characterization and potential utilization of municipal SolidWaste as renewable energy source: Lahore case study in Pakistan, Environ. Int. 134 (2020) 105291.

[50] X.X. HeY-R, S.G.-P. LiW-W, F.-F. Yan, H.-Q. Yu, H. Yuan, L.J. Wu, Enhanced electricity production from microbial fuel cells with plasma-modified carbon paper anode, Phys. Chem. Chem. Phys. 14 (2012) 9966–9971.

[51] R. Toczyłowska-Mamińska, K. Pielech-Przybylska, A. Sekrecka-Belniak, U. Dziekońska-Kubczak, Stimulation of electricity production in microbial fuel cells via regulation of syntrophic consortium development, Appl. Energy 271 (2020) 115184.

[52] Y.-F. Guan, F. Zhang, B.-C. Huang, H.-Q. Yu, Enhancing electricity generation of microbial fuel cell for wastewater treatment using nitrogen-doped carbon dots-supported carbon paper anode, J. Clean. Prod. 229 (2019) 412–419.

[53] L. Lan, J. Li, Q. Feng, L. Zhang, Q. Fu, X. Zhu, Q. Liao, Enhanced current production of the anode modified by microalgae derived nitrogen-rich biocarbon for microbial fuel cells, Int. J. Hydrog. Energy 45 (2020) 3833–3839.

[54] A. ElMekawy, S. Srikanth, S. Bajracharya, H.M. Hegab, P.S. Nigam, A. Singh, S.V. Mohan, D. Pant, Food and agricultural wastes as substrates for bioelectrochemical system (BES): the synchronized recovery of sustainable energy andWaste treatment, Food Res. Int. 73 (2015) 213–225.

[55] K. Dhamodharan, T. Konduru, M. Kannan, S.K. Malyan, Techno-economic feasibility and hurdles on agricultural waste management, in: Emerging Trends to Approaching Zero Waste, Elsevier, 2021, pp. 243–264, https://doi.org/10.1016/B978-0-323-85403-0.00011-6.

[56] A. Nandy, V. Kumar, P.P. Kundu, Effect of electric impulse for improved energy generation in mediatorless dual chamber microbial fuel cell through electroevolution of Escherichia coli, Biotechnol. Bioeng. 79 (2016) 796–801.

[57] B. Xiao, F. Yang, J. Liu, Enhancing simultaneous electricity production and reduction of sewage sludge in two-chamber MFC by aerobic sludge digestion and sludge pretreatments, J. Hazard. Mater. 189 (2011) 444–449.

[58] C.H. Ting, D.J. Lee, Production of hydrogen and methane from wastewater sludge using anaerobic fermentation, Int. J. Hydrog. Energy 32 (2007) 677–682.

[59] J. Parfitt, M. Barthel, S. Macnaughton, Food waste within food supply chains: quantification and potential for change to 2050, Philos. Trans. R. Soc. B 365 (2010) 3065–3081.

[60] J. Gustavsson, C. Cederberg, U. Sonesson, R. van Otterdijk, A. Meybeck, Global Food Losses and Food Waste, Food and Agriculture organization of the United Nations Publishing, Rome, 2011, pp. 1–23.

[61] S.K. Behera, J.M. Park, K.H. Kim, H.S. Park, Methane production from food waste leachate in laboratory-scale simulated landfill, Waste Manag. 30 (2010) 1502–1508.

[62] R.E. Speece, Anaerobic Biotechnology for Industrial Wastewater, Archae Press, Nashville (TN), 1996.

[63] S.S. Rikame, A.A. Mungray, A.K. Mungray, Electricity generation from acidogenic food waste leachate using dual chamber mediator less microbial fuel cell, Int. Biodeterior. Biodegradation 75 (2012) 131–137.

[64] L. He, P. Du, Y. Chen, H. Lu, X. Cheng, B. Chang, Z. Wang, Advances in microbial fuel cells for wastewater treatment, Renew. Sust. Energ. Rev. 71 (2017) 388–403.

[65] P.T. Kelly, Z. He, Understanding the application niche of microbial fuel cells in a cheese wastewater treatment process, Bioresour. Technol. 157 (2014) 154–160.

[66] S. Venkata Mohan, G. Mohanakrishna, G. Velvizhi, V.L. Babu, P.N. Sarma, Bio-catalyzed electrochemical treatment of real field DairyWastewater with simultaneous power generation, Biochem. Eng. J. 51 (2010) 32–39.

[67] M. Mahdi Mardanpour, M. Nasr Esfahany, T. Behzad, R. Sedaqatvand, Single chamber microbial fuel cell with spiral Anodefor dairy wastewater treatment, Biosens. Bioelectron. 38 (2012) 264–269.

[68] E. Elakkiya, M. Matheswaran, Comparison of anodic metabolisms in bioelectricity production during treatment of dairy wastewater in microbial fuel cell, Bioresour. Technol. 136 (2013) 407–412.

[69] J.A. Garcia-Nunez, N.E. Ramirez-Contreras, D.T. Rodriguez, E. Silva-Lora, C.S. Frear, C. Stockle, M. Garcia-Perez, Evolution of palm oil Mills into bio-refineries: literature review on current and potential uses of residual biomass and effluents, Resour. Conserv. Recycl. 110 (2016) 99–114.

[70] W.P.Q. Ng, H.L. Lam, F.Y. Ng, M. Kamal, J.H.E. Lim, Waste-to-wealth: green potential from palm biomass in Malaysia, J. Clean. Prod. 34 (2012) 57–65.

[71] E. Baranitharan, M.R. Khan, A. Yousuf, W.F.A. Teo, G.Y.A. Tan, C.K. Cheng, Enhanced power generation using controlled inoculum from palm oil mill effluent fed microbial fuel cell, Fuel 143 (2015) 72–79.

[72] S.E. Oh, B.E. Logan, Hydrogen and electricity production from food processing wastewater using fermentation and microbial fuel cell technologies, Water Res. 39 (2005) 4673–4682.

[73] A. Fogg, V. Gadhamshetty, D. Franco, J. Wilder, S. Agapi, S. Komisar, Can a microbial fuel cell resist the oxidation of tomato pomace? J. Power Sources 279 (2015) 781–790.

[74] M. Lu, S. Chen, S. Babanova, S. Phadke, M. Salvacion, A. Mirhosseini, S. Chan, K. Carpenter, R. Cortese, O. Bretschger, Long-term performance of a 20-L continuous flow microbial fuel cell for treatment of brewery wastewater, J. Power Source 356 (2017) 274–287.

[75] K. Scott, G.A. Rimbu, K.P. Katuri, K.K. Prasad, I.M. Head, Application of modified carbon anodes in microbial fuel cells, Process. Saf. Environ. Prot. 85 (2007) 481–488.

[76] P. Cheng, H.Z. Zhao, B. Zhao, J.R. Ni, Pilot treatment of wastewater from Dioscorea zingiberensis CH-Wright production by anaerobic digestion combined with a biological aerated filter, Bioresour. Technol. 100 (2009) 2918–2925.

[77] H. Li, J.R. Ni, W. Liu, Y.L. Zhu, Cleaner production alternatives for saponin industry by recycling starch, Resour. Conserv. Recycl. 54 (2010) 1145–1151.

[78] R. Liu, C. Gao, Y. Zhao, A. Wang, S. Lu, M. Wang, et al., Biological treatment of steroidal drug industrial effluent and electricity generation in the microbial fuel cells, Bioresour. Technol. 123 (2012) 86–91.

[79] H. Liu, B.E. Logan, Electricity generation using an air cathode single chamber microbial fuel cell in the presence and absence of a proton exchange membrane, Environ. Sci. Technol. 38 (2004) 4040–4046.

[80] H.C. Tao, T. Lei, G. Shi, X.N. Sun, X.Y. Wei, L.J. Zhang, et al., Removal of heavy metals from fly ash leachate using combined bioelectrochemical systems and electrolysis, J. Hazard. Mater. 264 (2014) 1–7.

[81] R.D. Cusick, P.D. Kiely, B.E. Logan, A monetary comparison of energy recovered from microbial fuel cells and microbial electrolysis cells fed winery or domestic wastewater, Int. J. Hydrog. Energy 35 (2010) 8855–8861.

[82] Y. Zhang, C. Sun, X. Liu, W. Han, Y. Dong, Y. Li, Electricity production from molasses wastewater in two-chamber microbial fuel cell, Water Sci. Technol. 68 (2013) 494–498.

[83] B. Zhang, H. Zhao, S. Zhou, C. Shi, C. Wang, J. Ni, A novel UASB–MFC–BAF integrated system for high strength molasses wastewater treatment and bioelectricity generation, Bioresour. Technol. 100 (2009) 5687–5693.

[84] G. Mohanakrishna, S. Venkata Mohan, P.N. Sarma, Bio-electrochemical treatment of DistilleryWastewater in microbial fuel cell facilitating decolorization and desalination along with power generation, J. Hazard. Mater. 177 (2010) 487–494.

[85] S. Sevda, X. Dominguez-Benetton, K. Vanbroekhoven, H. DeWever, T.R. Sreekrishnan, D. Pant, High Strength wastewater treatment accompanied by power generation using air cathode microbial fuel cell, Appl. Energy 105 (2013) 194–206.

[86] S.H.A. Hassan, S.M.F. Gad El-Rab, M. Rahimnejad, M. Ghasemi, J.-H. Joo, Y. Sik-Ok, I.S. Kim, S.-E. Oh, Electricity Generation from Rice straw using a microbial fuel cell, Int. J. Hydrog. Energy 39 (2014) 9490–9496.

[87] B. Cercado-Quezada, M. Delia, A. Bergel, Testing various food-industry wastes for electricity production in microbial fuel cell, Bioresour. Technol. 101 (2010) 2748–2754.

[88] M. Javed, M.A. Nisar, M. Ahmad, B. Muneer, Production of bioelectricity from VegetableWaste extract by designing a U-shaped microbial fuel cell, Pak. J. Zool. 49 (2017) 711–716.

[89] S. Venkata Mohan, G. Mohanakrishna, P.N. Sarma, Composite vegetable waste as renewable resource for bioelectricity generation through non-catalyzed open-air cathode microbial fuel cell, Bioresour. Technol. 101 (2010) 970–976.

[90] G. Mohanakrishna, S. Venkata Mohan, P.N. Sarma, Utilizing acid-rich effluents of fermentative hydrogen production process as substrate for harnessing bioelectricity: an integrative approach, Int. J. Hydrog. Energy 35 (2010) 3440–3449.

[91] S.A. Patil, V.P. Surakasi, S. Koul, S. Ijmulwar, A. Vivek, Y.S. Shouche, B.P. Kapadnis, Electricity generation using chocolate industry wastewater and its treatment in activated sludge based microbial fuel cell and analysis of developed microbial Community in the Anode Chamber, Bioresour. Technol. 100 (2009) 5132–5139.

[92] S. Pandit, N. Savla, J.M. Sonawane, A.M. Sani, P.K. Gupta, A.S. Mathuriya, A.K. Rai, D.A. Jadhav, S.P. Jung, R. Prasad, Agricultural waste and wastewater as feedstock for bioelectricity generation using microbial fuel cells: recent advances, Fermentation 7 (2021) 169, https://doi.org/10.3390/fermentation7030169.

[93] P. Pandey, V.N. Shinde, R.L. Deopurkar, S.P. Kale, S.A. Patil, D. Pant, Recent advances in the use of different substrates in microbial fuel cells toward wastewater treatment and simultaneous energy recovery, Appl. Energy 168 (2016) 706–723. ISSN 0306-2619, https://doi.org/10.1016/j.apenergy.2016.01.056.

CHAPTER 9

Microbial contamination in municipal water: Potential sources, analytical methods and remediation strategies

Dolly Rani[a], Vineeta Rana[b], Annu Rani[a], Sandeep K. Malyan[c], Avneesh Kumar[d], Rahul Kumar Dhaka[e], and Anuj Rana[a]

[a]Department of Microbiology, College of Basic Sciences & Humanities, CCS Haryana Agricultural University, Hisar, India
[b]Department of Botany and Plant Physiology (Environmental Science), College of Basic Sciences & Humanities, CCS Haryana Agricultural University, Hisar, India
[c]Department of Environmental Studies, Dyal Singh Evening College, University of Delhi, New Delhi, India
[d]Department of Botany, Shri Ramasamy Memorial University, Gangtok, Sikkim, India
[e]Department of Chemistry, College of Basic Sciences & Humanities, CCS Haryana Agricultural University, Hisar, India

1 Introduction

Microbial contamination in water is a global health issue which needs a real-time attention. Irregular and unhygienic practices result in emergence of microbial pollutants (algae, bacteria, fungi, protozoa and virus) in water supply chain. Municipal committee and government water regulatory authorities monitor and remediate these pathogens, in a particular area in water treatment plants, storage facilities, reservoirs and water pipe networks. The major sources of pathogens in water are disposal of fecal, industrial effluents, hospitals and other wastes (manure, slurry, household wastes). The pathogens cause mild (fever, headache, diarrhea) to lethal (cholera, typhoid, polio, cancer) diseases after ingestion of contaminated water. Chemical coating of pipelines, inadequate filtration, incomplete disinfection, poor infrastructure and structural design of the system lead to the growth of microorganisms (such as *Salmonella, Campylobacter, Shigella, Escherichia coli O157:H7, Cryptosporidium, Giardia and Norovirus*) in water bodies [1]. According to WHO report (March 2022), globally billions of people (~2 billions) consume contaminated water resulting into millions of deaths (~ 4.9 million) every year. A huge fraction of economy goes into waste for treatment of severe diseases due to consumption of contaminated water. It is important to employ advanced detection techniques to measure the presence of pathogens and strategies to control them. Presence of pathogens in water can be detected by turbidity, color, odor measurement and taste of water. Molecular and chemical methods are used to confirm the presence of microbial pollutants in water and certify the safety of same for drinking purpose [2]. Further, advanced technologies (such as PCR and sequencing) are used for efficient and rapid monitoring of microbial

Algae Based Bioelectrochemical Systems for Carbon Sequestration, Carbon Storage, Bioremediation and Bioproduct Generation
https://doi.org/10.1016/B978-0-323-91023-1.00009-0

pollutants in water. PCR is a target-specific approach to detect specific pathogen with high sensitivity; however, it requires expertise to handle the instrument [2].

Water treatment methods include filtration, disinfection, chemical treatment (precipitation/coagulation/adsorption) and biological treatment (microbial beds and bioelectrochemical system including algae with microbial fuel cells). These methods are categorized as primary, secondary and tertiary. Primary treatment includes filtration to remove live and dead cells. Membrane filtration (microfiltration and ultrafiltration) is a sustainable approach for removal of pathogens from water [3]. WHO has prescribed membrane filtration for effective removal of pathogens (protozoa and bacteria) [4]. For example, protozoa (*Giardia* and *Cryptosporidium*), bacteria (0.1 μm) and viruses (0.001 μm) can be removed by membrane filtration [5]. Membrane filtration technique is commonly used in industry and domestic domains; it needs sustainable and economic membrane filters for wide application for water treatment. Secondary treatment comprises disinfection of pathogens by chlorination, ozonation and UV radiation [6,7]. The other methods include physical, chemical and biological treatments (trickling filter, algal ponds and aquatic plants) [8,9]. Water treatment processes are generally labor intensive, require large setup and a skilled work force to operate.

Water treatment can be done using phytoremediation which, are eco-friendly, cost-effective and sustainable. Plants remove pathogens via attachment to roots or filtration process [10]. Pathogenic microorganisms (such as *Enterococci, E. coli, Klebsiella pneumonia, P. aeruginosa, Staphylococcus aureus, Salmonella typhi* and *Clostridium perfringens*) have been removed using aquatic plants such as *Typha latifolia, Cyperus papyrus, Phragmites mauritianus, Pistia stratiotes, Lemna paucicostata, Spirodela polyrhiza* and *Eichhornia crassipes* [11]. Economic, environmental friendly and sustainable material-based approaches are a better alternative for chemical-based water remediation strategies. Herein, sources of water-borne pathogens, their detection and remediation technologies (specific to algae) including their advantages and limitations are briefly discussed.

2 Water-borne pathogenic microorganisms

Microorganisms such as bacteria, virus, algae and protozoans contaminate water and transmit to humans. These pathogenic microorganisms cause serious illness to human through different site of infection inside the body by enumerating itself and production of harmful toxins. Commonly detected water-borne pathogens and disease caused by them are herein (Table 1).

2.1 Bacteria

Some pathogenic bacteria (e.g., *E. coli, Legionella, Mycobacterium avium, Salmonella, Shigella* and *Vibrio cholera*) grow in minimal requirement of nutrient sources including water and cause severe diseases (gastroenteritis, dysentery, typhoid, shigellosis and cholera, respectively) in humans once consumed [1]. These pathogens also contaminate water by

Table 1 Water-borne pathogens, their site of infection and disease.

S. No	Water-borne pathogens	Site of infection	Disease
Bacteria			
1	*Salmonella typhi*	Intestine and bloodstream	Typhoid
2	*Vibrio cholera*	Intestine	Cholera
3	*Shigella* sp.	The lining of the intestine	Shigellosis
5	*Legionella*	Lungs	Legionellosis
6	*Mycobacterium marinum*	Skin and soft tissues	Pulmonary infection
7	*Campylobacter jejuni*	Intestine	Campylobacteriosis
8	*E. coli*	Intestine	Gastroenteritis, diarrhea
9	*Clostridium botulinum*	Intestine	Botulism
Viruses			
1	*Adenovirus*	Lining of eyes, respiratory tract, intestinal tract, urinary tract, and nervous system	Adenovirus infections
2	*Astrovirus*	Intestine	Gastroenteritis
3	*Hepatitis A and E* viruses	Fecal-oral route	Hepatitis A and E
4	*Rotavirus*	Intestine	Rotavirus infection
5	*Norovirus*	Intestine	Norovirus infection
6	*Enteroviruses*	Oral-fecal route	Enterovirus infection
Cyanobacteria			
1	*Cyanobacteria*	Intestinal, liver, brain	Organ failure and neurological disorder
2	*Anabaena*		
3	*Fischerella*		
4	*Gloeotrichia*		
5	*Nodularia*		
6	*Nostoc*		
7	*Oscillatoria*		
8	*Planktothrix*		
Protozoans			
1	*Giardia*	Gastrointestinal tract	Giardiasis
2	*Cryptosporidium parvum*		Cryptosporidiosis

secretion of *enterotoxin* (*E. coli*) [12,13], cytolethal distending toxin (*Salmonella*) [14], shiga toxin (*Shigella*) and cholera toxin (*Vibrio cholera*) [15]. Consumption of such contaminated water may cause diarrhea (gentle to extravagant), nausea, vomiting and fast depletion of body fluid and electrolytes, resulted to demise if not treated properly. According to the current data of WHO, more than 2 billion people around the world are using fecal contaminated water, which cause disease like diarrhea and around 485,000 deaths per year due to its consumption [16].

2.2 Viruses

Viruses are present in small number in drinking water because they need host to replicate; however, water-borne viruses are more resistant than other viruses as they do not have a lipid envelop which makes their survival easy in digestive tract of human [17]. Different viruses present in water are *Adenovirus*, *Astrovirus*, *Hepatitis A and E* viruses, *Rotavirus*, *Norovirus* and *Enteroviruses* and cause severe diseases like gastroenteritis, meningitis, hepatitis (A and E) and paralysis [18]. These viruses replicate inside the living cells and leads to presymptoms of disease such as diarrhea, nausea, vomiting and fever.

2.3 Algae

Presence of harmful algal blooms in drinking water is a growing concern worldwide. The rapid growth of blue green algae (Cyanobacteria) forms algal blooms and contaminate the water sources through production of harmful cyanotoxins. Cyanobacteria release different types of toxins like microcystins, saxitoxin, anatoxin, ciguatoxin, β-*N*-methylamino-L-alanine (BMAA), lipopolysaccharide (LPS) endotoxins, cylindrospermopsin, and nodularins [19]. Algae such as *Nostoc*, *Anabaena*, *Fischerella* and *Oscillatoria* can spread infection in human through contaminated water [20]. It can cause fever, diarrhea, headache, vomiting, gastrointestinal infection, kidney and liver damage and may leads to death in severe cases [21]. Algae can contaminate water bodies through agriculture waste by formation of algal blooms and release of toxins as by-product in the water. Two harmful algae strains *Microseira wollei* and *Phormidium* were found as contaminant in South Carolina lakes [22].

2.4 Protozoa

Protozoans can also contaminate the drinking water, through entering from sewage and fecal waste. *Giardia* and *Cryptosporidium* are the commonly detected protozoans in fresh water, which are responsible for diarrheal diseases giardiasis and cryptosporidiosis [23]. They can survive in unfavorable conditions by formation of cyst (dormant phase of protozoans). Ingestion of such contaminated water can cause diseases such as diarrhea, stomach cramps, nausea, fatigue, dehydration and headache [24].

3 Detection methods for waterborne biocontaminants

The detection of biological contaminants (algae, bacteria, fungi, virus and protozoa) in water with accuracy and sensitivity is important to establish the real situation of threat and formulate the guidelines for adoption of a suitable treatment process. The technique for the detection of biological contaminants must be rapid, sensitive and user-friendly [25]. The available methods for microbial contaminants detection directly detect the most probable contaminants, but they do not cover all the possibilities. Therefore, it is important to include complementary analytical methods. Methods for biological contaminants detection can be broadly classified as culture based (classical approach) and molecular methods. Techniques like microfluidic devices, spectroscopic approaches and biosensors have also been developed for the detection of pathogen for real-time monitoring. The selection of a suitable technique depends on the intended purpose of analysis (quantitative and qualitative). Despite the advancement of the techniques, the conventional culture-based methods are still in use for the determination of microbiological parameters in water [26]. The following sections deal with different analytical methods (Table 2) including their merit and demerits for biological pollutants detection in water.

Table 2 Comparative evaluation of analytical methods used to detect biological contaminants in water [26]

Analytical method	Merits	Demerits
Polymerase chain reaction method	• High detection rate, • High sensitivity	• Loss of amplification effectiveness due to presence of inhibitors
Emission, optical, luminescence method (light emission spectroscopy)	• Most commonly used technique	• Sensitivity alteration with temperature • It needs skillful supervision • Great care is needed during analysis due to illumination sensitivity
Biosensors	• Better sensitivity • No any sample pre-processing is required • Quick working time • Appropriate for in-situ observation	• Sensitive to situations • Lack of system consistency • Restricted transducer life span • Threat of bioreceptor leaks
Sensor placements approach	• High sensitivity event detection • Multiple water quality parameter measurements	• Comparatively high cost • Small datasets • Complex design analysis and optimization • High inaccuracy rate • Transmission time delay • High maintenance
Microfluidics sensors	• Label-free detection technique • Manageable and reasonable technique • Least sample volume • Fast response and • High detection sensitivity	• Exclusive setup and preprocessing of sample is required • Hand-operated sample collection

Continued

Table 2 Comparative evaluation of analytical methods used to detect biological contaminants in water—cont'd

Analytical method	Merits	Demerits
EIS and DIS spectroscopy	• Label-free detection tool • Simplicity and cost-effective	• Insufficient recognition sensitivity • Noncontinuous detection • Skilled supervision is required • Light sensitive
Light emission/luminescence	• Simple and fast detection of BOD/DOM • Easy separation phases • Dense and nondestructive technique	• Large-scale applications are costly • Preparation of sample is required • Narrow range of parameters for detections of water quality • Temperature sensitive
IR, MIR, NIR	• Economical and simple techniques • Quick detection speeds • Convenient and stable dimensions	• Elevated interference of water • Appearance of overlying • Absorption band along with overtones • Reduced signal strength
Discontinuous (sample-based) analysis	• Precise impurities detection • Good numerical data measurements	• Laborious process taking 48–96 h to complete • Sensitivity issues with low concentration • Narrow contamination detection

3.1 Culture-dependent method

Culture-dependent method is labor intensive, require long analysis time (usually 1–3 days for culture growth) and have low sensitivity. The most routinely used test is plate count to enumerate the bacteria. A wide range of microorganisms including bacteria, fungi, molds and yeast are analyzed simultaneously on a single or different culture media. This method helps in monitoring bacteriological quality of contaminated water by direct counting the microbial load in the treated and wastewater. This method can detect only culturable viable cells present in water [27]. The other pathogens which cannot grow on the defined media need to be detected through culture independent methods such as PCR, sequencing and other advanced techniques.

3.2 Turbidity based method

It is a physical approach to assess the risk associated with microbiological agents in an easy and feasible way. However, there is a lack of direct relation between turbidity and water-borne pathogen density; still it is commonly used in wastewater treatment plants. This parameter is suitable to detect water biological contaminants as suspended particles. Turbidity in water allows deposition of particles promoting the formation and growth of biofilms for potential pathogens like *Legionella*, *Pseudomonas*, *Aeromonas*, *Mycobacterium* and reduce the efficiency of disinfectant. The use of turbidity to continuously assess and monitor effluent quality is a primitive method for detection of pathogen load [28]. In addition, quantification of organic carbon with turbidity can provide better accuracy for measurement of bacterial contamination [28].

3.3 Multiple tube fermentation technique (MTF)

This technique (MTF) is a standard laboratory method used for the detection of microbial activity in water sample. This technique includes three stages: presumptive, confirmed and completed. Presumptive stage consists of a series of tube incubation process leading to the formation of gas, which indicates positive presumptive test. Further, the sample was grown on culture medium for confirmation [26]. Detection of *E. coli* and other gas-producing bacteria in water was done by MTF with high detection rate. Additionally, MTF technique showed promising results in identification of yeast.

3.4 Membrane filtration (MF) method

Membrane filtration based method is suitable for isolation and elimination of microorganisms in large sample volume as compared to the previous MTF technique. MF method is recognized by United States Environmental Protection Agency (USEPA) and United Nations Environment Program (UNEP/WHO) for biological contaminants detection in potable water [5]. A direct count of bacterial colonies could be obtained by incubating the membrane filter on selective media (solid or liquid) at optimum temperature. Further, this technique also finds its use in retaining virus. Advancement of technique includes simultaneous use of microfiltration and ultrafiltration membrane filters, thus improving the efficiency of process.

3.5 Algal biomass method

Algal biomass can be used for detection of water contaminant by analyzing photosynthetic rate, where energy movement, sound waves and expansion of molecules are measured. The frequency and wavelength of these waves are analyzed using microphones. It helps to monitor the algae health as per type and concentration of water contaminants. This method is comparatively accurate, fast and requires less instrumentation [29].

3.6 Immunoassay-based methods

Immunoassay-based methods depend on interaction of designed antibodies with specific antigen present on the surface of pathogen. These methods are used for the identification of pathogens such as bacteria and viruses. These methods involve lateral flow technology having the advantage of being short analysis time (15–30 min) [25]. Such technologies can be used for rapid screening of pathogens qualitative as well as quantitative. The technologies based on these methods do not require special skills.

3.7 Fluorescence in situ hybridization (FISH)

FISH technique enables the detection, identification, localization and enumeration of microorganisms. The cellular components in targeted microorganisms are visualized using fluorescence probe. These techniques have wide applications in cytogenetic, microbiology and genetic diagnostics [26]. This technique is also used for detection of bacteria in living tissues as well as in aquatic environmental samples using DNA probe (16s rRNA gene). Execution of FISH requires expertise in probe designing and selection of fluorophore. Detection of microbial community is limited by FISH technique as it produces low fluorescence signal. This limitation can be overcome by using multilabeled FISH technique where fluorescence signals can improve for simultaneous microbial detection in contaminated water.

3.8 Molecular methods

3.8.1 Polymerase chain reaction (PCR)

PCR is used for rapid detection of pathogens in water by amplifying a single copy or few copies of targeted genetic material in vitro. Cloning-based replication of DNA sequence takes days or week to complete, while the amplification of DNA sequence using PCR takes place very quickly (within hours). PCR possesses high level of amplification with high sensitivity and reliability, which makes it useful technique for commercial application including water quality control [25]. Recently, several developments have been made in advancement of PCR techniques (qPCR, RT-qPCR, loop-mediated isothermal amplification, strand displacement amplification). The qPCR is an automatized and sensitive technique for simultaneous amplification and quantification of target microorganism [26].

3.8.2 Sequencing

Sequencing is an advanced technique, which can be used for the detection of pathogens based on their genetic material. There are many sequencing approaches from classical to modern for the analysis of genomics. Most advanced, next-generation sequencing platform can be used for sequencing of millions of reads of target sequence at a single time. There are some real-time and comparative economic portable sequencing devices such as minION for sequencing of genetic material of different microorganisms. Genomic libraries can also be prepared using a minION field sequencing database for real-time monitoring of specific microorganisms [25]. Unlike traditional sequencing techniques, long read nanopores gene sequencing allows for a characterization of native RNA or DNA without fragmentation or amplification, which streamlined the analysis and removes potential source of mismatching [25].

3.8.3 Flow cytometry and fluorimetry

Cytometry follows optical principle (light scattering and fluorescence) to detect microorganism at cellular level and gives information regarding individual cell in heterogeneous population. This technique is frequently used off-site to assess microbe ecology [28]. Fluorescent probes are being used to analyze bacterial and algal cell while, some autotrophic and heterotrophic bacteria use their natural pigments as fluorescence. It is a multiparametric analysis, which determine the total cell count (TCC) and bacterial viability (e.g., SYTO and SYTOX nucleic acid stains) [30] simultaneously. Recently, flow cytometry extends its application to assess ground water quality in terms of bacterial dynamics, making this technique acceptable significantly. The method involves automated water sampling at 15-min intervals and analyzing bacterial population via flow cytometry. This analysis can be done consecutively for 14 days. Flow cytometry systems like CytoSense, CytoBuoy and CytoSub are fully automated and used to check the toxicity on phytoplankton. Automated fluorimetry connected with online database provides a continuous evaluation of bacterial contaminated water. Specifically, tryptophan like fluorescence (TLF) measures total organic carbon, biological oxygen demand (BOD) and chemical oxygen demand (COD) of a contaminated water body. Contaminated water with fecal *Coliforms* species shows a gradual increase in TLF value with the proliferation of microorganism in water [30].

3.9 Biosensors

Biosensor-based devices are being used for pathogen detection. Biosensors measure molecular signals by applying specific biorecognition patterns (enzymes, cell, antibodies and nucleic acids) which are further integrated with electrical interferences via transducer platforms to obtain signals. A number of bacterial species like *E. coli, Aliivibrio fischeri* and *Photobacterium phosphoreum* are examined in biomonitoring of contaminated water sample. *Pseudomonas* and *Flavobacterium* are generally used as biosensors for detection of biological contaminant in solid and liquid [30]. Biosensors exhibit the potential for detection of *E. coli* using fluorescence and chemiluminescence probes. Biosensing system cooperated with nanoparticles of gold, graphene or quantum dots contributes to the scientific advancement in this field [28]. There are several advantages associated with biosensors including high limit of detection, reproducibility, reliability, fast response, easy to operate, high accuracy and economic efficiency.

3.9.1 Carbon nanotube-based sensors

Carbon-based nanosensor is an advance technology for detection and removal of bacteria, virus, pathogens and other biological contaminants from water. These sensors efficiently detect the type and concentration of contaminants by producing electrical signals. These sensors are made of ultrathin material, possessing durability and high conductivity to heat and electricity [29].

3.9.2 Nanobiosensors

Nanoparticles of gold, silver, silicon and zinc are used for detection and removal of pathogenic bacteria from wastewater. Global acceptance of nanoparticles in wastewater treatment is due to their low cost, safe to use and fast analysis. Nanomaterial-based sensors and detectors are highly selective, rapid, reliable and sensitive, which makes them better toward sensing and detection technique against pathogens like bacteria and viruses [31]. For example, gold nanoparticle combined with *Salmonella* antibody can be used to detect *S. typhimurium* (10^4 cells/mL) using colorimetric assay in drinking water. Further, sensitivity of nanoparticles is dependent on its shape and size. Nanozeolite (serves as a carrier silver nanoparticles)-based detection technique is novel for real-time monitoring of viruses and other biological contaminants containing large protein particles. It consists of nanofluidic channel along with optical interferometry causing scattering of light by single virus (producing characteristic signals of a particular virus) which is detected by heterodyne interferometer [31].

3.10 Scope for advancement in detection technology

Advancement in technologies enhances the sensitivity and accuracy for detection of water biopollutants but still associated with some demerits and eventually affects the results. The technologies for biological contaminants detection may be improved in terms of low cost and easy in operation. The technologies may be advanced for the commercialization and must accurately determine the biological contaminants without requirement of sophisticated instruments and trained persons.

4 Drinking water microbial pollutant remediation strategies

4.1 Physical methods

Microbial contaminants in water can be removed by trapping the microorganisms using different size of filters and their survival can be restricted through disturbing the cellular morphology using different physical treatment. Some preliminary and eco-friendly commonly used methods to control the microbial growth in water are described below.

4.1.1 Filtration

Filtration is a size-dependent process in which microbial contaminants can be trapped through different size of filters, for example, 1 μm (protozoans), 0.1–0.5 μm (blue-green algae and bacteria) and 0.01–0.001 μm (virus) filters. Many filtration processes are available such as sand filtration, microfiltration, ultrafiltration through reverse osmosis and nanofiltration. Ultrafiltration has pore size less than 0.01 μm, and it can be helpful for removal of viral contaminants [32]. Technology improvement has been done through development and application of nanofilters to trap the microorganisms in water. The size of nanofilters varies from 10 to 100 nm for enhanced absorption capacity for removal of microbial contaminants from drinking water [33]. Nanofilters have been effectively used to remove microbial contaminants in water, for example, silver-based nanofilters (for *E. coli*) and cellulose-based nanofilters (for viral pathogens like HIV, bovine diarrhea virus and reovirus) [34]. The nanotechnology has many advantages over classical filtration approaches; however, there are challenges to develop ecofriendly and cost-effective nanofilters for establishment of sustainable approach. There are some limitations to use these filters as it requires proper cleaning, washing, time consumption and energy-dependent processes to pass the water through the filters. Integrated use of filtration techniques has successfully removed the microbial contaminates in drinking water.

4.1.2 Radiations

Ultraviolet (UV) radiations have destructive effect on the growth of all microorganisms by forming pyrimidine dimers in their genetic material, which results to interfere with transcription and translation [35,36]. Almost all pathogens can be killed using UV range from 200 to 400 nm with a fluence of 2–10 mJ/cm^2 (*S. typhi*), 1–5 mJ/cm^2 (*Shigella dysenteriae*), 8–306 mJ/cm^2 (*Adenovirus*), 5–28 mJ/cm^2 (Hepatitis A), 1.5 W/cm^2 (*Chlamydomonas reinhardtii*), 0.5–6.1 mJ/cm^2 (*C. parvum*) and 1.5–11 mJ/cm^2 (*Giardia muris*) [37,38]. Ultrasonic radiations can pass through the medium, penetrate and damage the microbial cell structure and kill pathogens at frequency (42 kHz for 150 s) [39,40].

4.2 Chemical disinfectants

Water treatment has been done for removal/kills microbial contaminants by using different chemicals such as chlorine, ozone and iodine. Disinfection using chlorine is proven to be an effective approach, and it is used in worldwide to disinfect public water [41]. Water treatment with chlorine ≤ 1 mg/L for ~ 30 m can kill most of the harmful bacteria [42]. However, viruses and cysts of protozoa are resistant to chlorine. Ozone has been used as a powerful oxidizing agent to kill microorganisms by oxidizing and rupturing the membrane, which results to kill the pathogens. Ozone is considered a good disinfectant as no traces of ozone leftover in water after treatment. However, ozonation has some side effects such as irritant to skin, eyes, respiratory tract as well as mucosal membrane. Still some gram positive bacteria and viruses are resistant to ozone treatment [43]. Iodine is a simple, cost-effective and efficient approach to kill bacteria in few minutes, but it takes hours to kill protozoans [44]. However, it is not recommended for ingestion [45]. Huang and co-workers evaluated the effect of two agents (powdered activated carbon and chlorine) for the removal of pathogenic cyanobacteria (*Microcystis aeruginosa* and *Pseudoanabaena* sp.) with enhanced efficacy from wastewater [46].

4.3 Biological water treatment

Biological treatment methods are environment friendly approaches to trap or kill pathogens in water. Biological treatment methods are secondary treatment methods which form biofilms on activated surfaces (trickling filters) to trap and kill pathogens via oxidation–reduction reactions. They are also used as activated sludge via enrichment of microbial population to treat contaminated water. Oxidation system can kill 90% of microbial pathogens [47]. Membrane-based bioreactors are used to remove microbial contaminants in industrial and municipal water [48]. Under this procedure, microorganisms also produce some phenolic compounds which will help to kill pathogens [49]. These bioreactors are used with integration of adsorbent filters such as granular activated carbon, ultrafilters and vegetated filter strips for efficient removal of water-borne pathogens [50]. Bacterial colonization takes place over granular activated carbon to form biofilm which acts as a mass of slime matrix containing biopolymers (polysaccharides) to trap pathogens [51]. Aquatic plants have been used to treat contaminated water with 90% removal efficiency against tested pathogenic microorganisms (*Enterococci, E. coli, Klebsiella pneumonia, Pseudomonas aeruginosa, C. perfringens, S. aureus* and *Salmonella*) [11]. Aquatic plants (*T. latifolia, C. papyrus, P. mauritianus, P. stratiotes, L. paucicostata*) trap the pathogen with their roots and deprive them for nutrient availability which leads to kill the pathogens in the water. Delanka-Pedige and co-workers evaluated the capability of *Galdieria sulphuraria* for the reduction of pathogens. The results showed 3.3 log units reduction of total coliform in the effluent (2.3×10^7 CFU/100mL). Further qPCR analysis confirmed the removal of *Enterococcus faecalis*, *E. coli* and 98% removal of total bacteria [52].

4.3.1 Algal-based bioelectrochemical system for removal of microbial contamination

Bioelectrochemical system is being used for water management to treat effluent from different sources such as sewage [53], industry [54], textile [55], agro-waste [56], pharmaceutical [57] and petrochemical [58]. Bioelectrochemical system is an emerging and inventive approach for treatment of domestic water [59], ground water [60] and industrial water [61] by bioconversion of organic substrates using redox reactions.

Algae has been used as a promising agent in bioelectrochemical system as it leads to the simultaneous removal of organic matter and reduced CO_2 emission, which makes it more attractive and advantageous to traditional biological water treatment systems. Application of microalgae as biocathode is a cost-effective and environmental friendly approach as it reduces the expenditures on expensive catalysts (platinum), chemicals and buffers. In the presence of sunlight, algae convert CO_2 to organic matter and generate oxygen and biomass; while during dark stage, they consume oxygen and yield energy by direct oxidation of the organic matter produced in light reaction [62]. Xiao and co-workers developed a photoelectron-biochemical system comprised of algal bioreactor and microbial fuel cell

(MFC) to reduce organic content (92% COD removal) from water using MFC and nutrient depletion (98% ammonium nitrogen and 82% phosphate removal) by algal bioreactor [63].

Integrated approaches toward green technology, for example, microbial fuel cell electrochemically active bacteria (MFC EAB) with algae oxidize organic matter to generate electrons and produced CO_2, where algae worked as electron acceptor and donor under dark and light conditions, respectively. During this process, it develops syntrophic interaction with electrochemically active bacteria [64]. Such integrated technologies can be used for effective treatment of wastewater, with addition to reduced gas emission, solid and sludge production [65]. Bazdar and co-workers (2018) have developed a photosynthetic microalgae MFC using *Chlorella vulgaris* for removal of COD (5.47%) from wastewater [66]. Integrated use of membrane photo-bioreactor using algae and MFC removed (92%–97%) soluble chemical oxygen demand (COD) and nearly complete removal of ammonia from wastewater. Such systems can be used with better efficiency for water treatment at industrial scale [67].

4.4 Hybrid technologies

A single technology cannot remove all types of microbial contaminants. Therefore, integrated use of primary and secondary treatment approaches is used. Technologies have been improved for water treatment in terms of efficiency and acceptability by combining two or more separation steps in an integrated manner (hybrid technologies) [68]. For example, reverse osmosis (RO) units are designed with filter, ion-exchange technology and UV to remove pathogens in any form and used widely for domestic drinking water treatment [69]. Recently, granular media are used in filtration to deal with problems like membrane damage, fouling due to acids and/or chlorine [69].

5 Conclusion and recommendations

Fresh water on the earth cannot be used directly for drinking purpose due to presence of pathogens including other pollutants. Biological contaminant of drinking water is an emerging global problem. In this chapter, we have overviewed the detection techniques for microbial contamination in water including their feasibility and limitations. Traditional culture-based approaches and molecular methods for the identification of specific pathogens have been included in the discussion. These pathogens can be trapped, removed or killed by using different treatment methods, namely physical (filtration, radiation), chemical (chlorination, iodization, ozonation) and biological (biological oxidation and reduction through microorganisms, adsorption via aquatic plants and algal biomass and/or bioadsorbent thereof). Sustainable materials (such as biomass based (nanomaterials) may be employed to enhance efficiency and removal kinetics of pathogens. The efficiency of biological treatment methods can be improved using integrated approaches as a single method cannot remove all the pathogens completely. A suitable support/carrier material may be incorporated to enhance the efficacy and stability of microorganisms [68].

Acknowledgment

The authors (D.R., V.R. and A.R.) are thankful to the university for providing the research fellowship. R.K.D. and A.R. thank to Chaudhary Charan Singh Haryana Agricultural University, Hisar, for providing necessary research facilities.

References

[1] N.J. Ashbolt, Microbial contamination of drinking water and human health from community water systems, Curr. Environ. Health Rep. 2 (2015) 95–106.
[2] F.Y. Ramírez-Castillo, A. Loera-Muro, M. Jacques, P. Garneau, F.J. Avelar-González, J. Harel, A.L. Guerrero-Barrera, Waterborne pathogens: detection methods and challenges, Pathogens 4 (2015) 307–334.
[3] F.I. Hai, T. Riley, S. Shawkat, S.F. Magram, K. Yamamoto, Removal of pathogens by membrane bioreactors: a review of the mechanisms, influencing factors and reduction in chemical disinfectant dosing, Water 6 (2014) 3603–3630.
[4] C. Charcosset, Classical and recent developments of membrane processes for desalination and natural water treatment, Membranes 12 (2022) 267.
[5] E. Obotey Ezugbe, S. Rathilal, Membrane technologies in wastewater treatment: a review, Membranes 10 (2020) 89.
[6] K. Johnson, M.A. Kumar, P. Ponmurugan, B.M. Gananamangai, Ultraviolet radiation and its germicidal effect in drinking water purification, J. Phytol. 2 (2010) 12–19.
[7] J. Gomes, A. Matos, M. Gmurek, R.M. Quinta-Ferreira, R.C. Martins, Ozone and photocatalytic processes for pathogens removal from water: a review, Catalysts 9 (2019) 46.
[8] P. Choudhary, S.K. Prajapati, P. Kumar, A. Malik, K.K. Pant, Development and performance evaluation of an algal biofilm reactor for treatment of multiple wastewaters and characterization of biomass for diverse applications, Bioresour. Technol. 224 (2017) 276–284.
[9] S. Sharma, A. Bhattacharya, Drinking water contamination and treatment techniques, Appl. Water Sci. 7 (2017) 1043–1067.
[10] S.I. Alexandros, C.S. Akratos, Removal of pathogenic bacteria in constructed wetlands: mechanisms and efficiency, in: Phytoremediation, Springer, 2016, pp. 327–346.
[11] B. Dhir, Effective control of waterborne pathogens by aquatic plants, in: Waterborne Pathogens, Elsevier, 2020, pp. 339–361.
[12] G.A. Brito, C. Alcantara, B.A. Carneiro-Filho, R.L. Guerrant, Pathophysiology and impact of enteric bacterial and protozoal infections: new approaches to therapy, Chemotherapy 51 (2005) 23–35.
[13] J.P. Nataro, J.B. Kaper, Diarrheagenic *Escherichia coli*, Clin. Microbiol. Rev. 11 (1998) 142–201.
[14] S.K. Niyogi, Shigellosis, J. Microbiol. 43 (2005) 133–143.
[15] J. Theron, T. Cloete, Emerging waterborne infections: contributing factors, agents, and detection tools, Crit. Rev. Microbiol. 28 (2002) 1–26.
[16] W.H. Organization, Water, Sanitation, Hygiene and Health: A Primer for Health Professionals, World Health Organization, 2019.
[17] A. Pinon, M. Vialette, Survival of viruses in water, Intervirology 61 (2018) 214–222.
[18] P. Shankar, J. Mishra, S. Singh, Hepatitis A and E in Potable Water: A Threat to Health, Water and Health, Springer, 2014, pp. 29–51.

[19] X. He, Y.-L. Liu, A. Conklin, J. Westrick, L.K. Weavers, D.D. Dionysiou, J.J. Lenhart, P.J. Mouser, D. Szlag, H.W. Walker, Toxic cyanobacteria and drinking water: impacts, detection, and treatment, Harmful Algae 54 (2016) 174–193.

[20] M. Figgatt, N. Muscatiello, L. Wilson, D. Dziewulski, Harmful algal bloom–associated illness surveillance: lessons from reported hospital visits in New York, 2008–2014, Am. J. Public Health 106 (2016) 440–442.

[21] R. Wood, Acute animal and human poisonings from cyanotoxin exposure—a review of the literature, Environ. Int. 91 (2016) 276–282.

[22] M.T. Aziz, C.O. Granger, D.C. Westerman, S.P. Putnam, J.L. Ferry, S.D. Richardson, *Microseira wollei* and *Phormidium* algae more than doubles DBP concentrations and calculated toxicity in drinking water, Water Res. 216 (2022) 118316.

[23] S. Baldursson, P. Karanis, Waterborne transmission of protozoan parasites: review of worldwide outbreaks–an update 2004–2010, Water Res. 45 (2011) 6603–6614.

[24] A. Omarova, K. Tussupova, R. Berndtsson, M. Kalishev, K. Sharapatova, Protozoan parasites in drinking water: a system approach for improved water, sanitation and hygiene in developing countries, Int. J. Environ. Res. Public Health 15 (2018) 495.

[25] M.R. Coelho, M. Batlle Ribas, M.F. Coimbra, Review of Technologies for the Rapid Detection of Chemical and Biological Contaminants in Drinking Water, ERNCIP Chemical and Biological Risks to Drinking Water Thematic Group, 2020.

[26] S.N. Zulkifli, H.A. Rahim, W.-J. Lau, Detection of contaminants in water supply: a review on state-of-the-art monitoring technologies and their applications, Sensors Actuators B Chem. 255 (2018) 2657–2689.

[27] T. Manickum, Total colony counts (TCC) by flow cytometry (FCM) should replace the heterotrophic plate count (HPC) test for bacteriological enumeration of water-some recent developments in flow cytometry: a review, Hydrol. Curr. Res. 11 (2020).

[28] A.-V. Jung, P. Le Cann, B. Roig, O. Thomas, E. Baurès, M.-F. Thomas, Microbial contamination detection in water resources: interest of current optical methods, trends and needs in the context of climate change, Int. J. Environ. Res. Public Health 11 (2014) 4292–4310.

[29] M. Peyravi, M. Jahanshahi, H. Tourani, Analytical methods of water pollutants detection, in: Inorganic Pollutants in Water, Elsevier, 2020, pp. 97–113.

[30] D. Wlodkowic, T.M. Karpiński, Live-cell systems in real-time biomonitoring of water pollution: practical considerations and future perspectives, Sensors 21 (2021) 7028.

[31] A. Anand, U. Rajchakit, V. Sarojini, Detection and removal of biological contaminants in water: the role of nanotechnology, in: Nanomaterials for the Detection and Removal of Wastewater Pollutants, Elsevier, 2020, pp. 69–110.

[32] B. Van der Bruggen, C. Vandecasteele, Removal of pollutants from surface water and groundwater by nanofiltration: overview of possible applications in the drinking water industry, Environ. Pollut. 122 (2003) 435–445.

[33] R. Singh, R. Bhadouria, P. Singh, A. Kumar, S. Pandey, V.K. Singh, Nanofiltration technology for removal of pathogens present in drinking water, in: Waterborne Pathogens, Elsevier, 2020, pp. 463–489.

[34] V. Mangayarkarasi, A. Prema, N. Rao, Silver nanomembrane and ceramic silver nanofilter for effective removal of water borne diarrhoegenic *Escherichia coli*, Indian J. Sci. Technol. 5 (2012) 2029–2034.

[35] N. Goosen, G.F. Moolenaar, Repair of UV damage in bacteria, DNA Repair 7 (2008) 353–379.

[36] T.D. Cutler, J.J. Zimmerman, Ultraviolet irradiation and the mechanisms underlying its inactivation of infectious agents, Anim. Health Res. Rev. 12 (2011) 15–23.

[37] W. Hijnen, E. Beerendonk, G.J. Medema, Inactivation credit of UV radiation for viruses, bacteria and protozoan (oo) cysts in water: a review, Water Res. 40 (2006) 3–22.

[38] T. Sydney, J.-A. Marshall-Thompson, R.V. Kapoore, S. Vaidyanathan, J. Pandhal, J.P.A. Fairclough, The effect of high-intensity ultraviolet light to elicit microalgal cell lysis and enhance lipid extraction, Metabolites 8 (2018) 65.

[39] A. Mahvi, M. Dehghani, Evaluation of ultrasonic technology in removal of algae from surface waters, Pak. J. Biol. Sci. 8 (2005) 1457–1459.

[40] M.H. Dehghani, Effectiveness of ultrasound on the destruction of *E. coli*, Am. J. Environ. Sci. 1 (2005) 187–189.

[41] M. Kerwick, D. Holt, M. Kerwick, S. Reddy, A. Chamberlain, A methodology for the evaluation of disinfection technologies, J. Water Health 3 (2005) 393–404.

[42] D. Ghernaout, Water treatment chlorination: an updated mechanistic insight review, Chem. Res. J. 2 (2017) 125–138.

[43] F. Am Water Works Res, B. Langlais, D.A. Reckhow, D.R. Brink, Ozone in Water Treatment: Application and Engineering, CRC Press, 1991.

[44] P.R. Hove, D. Mobley, F. Magunda, D.R. Call, Deploying elemental iodine in a vapor form to disinfect water and to clear biofilms, Int. J. Environ. Res. Public Health 17 (2020) 3489.

[45] H. Backer, J. Hollowell, Use of iodine for water disinfection: iodine toxicity and maximum recommended dose, Environ. Health Perspect. 108 (2000) 679–684.

[46] X. Huang, H. Liang, Y. Yu, B. Shi, The enhanced treatment of algae-laden water by combination of powdered activated carbon and chlorine, Sep. Purif. Technol. 280 (2022) 119922.

[47] N. Abdel-Raouf, A. Al-Homaidan, I. Ibraheem, Microalgae and wastewater treatment, Saudi J. Biol. Sci. 19 (2012) 257–275.

[48] S. Deowan, S. Bouhadjar, J. Hoinkis, Membrane Bioreactors for Water Treatment, in: Advances in Membrane Technologies for Water Treatment, Elsevier, 2015, pp. 155–184.

[49] P. Le-Clech, Membrane bioreactors and their uses in wastewater treatments, Appl. Microbiol. Biotechnol. 88 (2010) 1253–1260.

[50] A. Guber, A. Yakirevich, A. Sadeghi, Y. Pachepsky, D. Shelton, Uncertainty evaluation of coliform bacteria removal from vegetated filter strip under overland flow condition, J. Environ. Qual. 38 (2009) 1636–1644.

[51] D.R. Simpson, Biofilm processes in biologically active carbon water purification, Water Res. 42 (2008) 2839–2848.

[52] H.M.K. Delanka-Pedige, S.P. Munasinghe-Arachchige, J. Cornelius, S.M. Henkanatte-Gedera, D. Tchinda, Y. Zhang, N. Nirmalakhandan, Pathogen reduction in an algal-based wastewater treatment system employing *Galdieria sulphuraria*, Algal Res. 39 (2019) 101423.

[53] J. Kim, K. Kim, H. Ye, E. Lee, C. Shin, P.L. McCarty, J. Bae, Anaerobic fluidized bed membrane bioreactor for wastewater treatment, Environ. Sci. Technol. 45 (2011) 576–581.

[54] Y. Dong, Y. Qu, W. He, Y. Du, J. Liu, X. Han, Y. Feng, A 90-liter stackable baffled microbial fuel cell for brewery wastewater treatment based on energy self-sufficient mode, Bioresour. Technol. 195 (2015) 66–72.

[55] S. Cheng, P. Kiely, B.E. Logan, Pre-acclimation of a wastewater inoculum to cellulose in an aqueous–cathode MEC improves power generation in air–cathode MFCs, Bioresour. Technol. 102 (2011) 367–371.

[56] P. Pushkar, A.K. Mungray, Real textile and domestic wastewater treatment by novel cross-linked microbial fuel cell (CMFC) reactor, Desalin. Water Treat. 57 (2016) 6747–6760.

[57] D. Molognoni, S. Chiarolla, D. Cecconet, A. Callegari, A.G. Capodaglio, Industrial wastewater treatment with a bioelectrochemical process: assessment of depuration efficiency and energy production, Water Sci. Technol. 77 (2018) 134–144.

[58] V. Ashwaniy, M. Perumalsamy, Reduction of organic compounds in petro-chemical industry effluent and desalination using *Scenedesmus abundans* algal microbial desalination cell, J. Environ. Chem. Eng. 5 (2017) 5961–5967.

[59] A. Capodaglio, D. Molognoni, E. Dallago, A. Liberale, R. Cella, P. Longoni, L. Pantaleoni, Microbial fuel cells for direct electrical energy recovery from urban wastewaters, Sci. World J. 2013 (2013).

[60] D. Cecconet, A. Callegari, P. Hlavínek, A.G. Capodaglio, Membrane bioreactors for sustainable, fit-for-purpose greywater treatment: a critical review, Clean Techn. Environ. Policy 21 (2019) 745–762.

[61] A. Callegari, D. Cecconet, D. Molognoni, A.G. Capodaglio, Sustainable processing of dairy wastewater: long-term pilot application of a bio-electrochemical system, J. Clean. Prod. 189 (2018) 563–569.

[62] A. Commault, G. Lear, P. Novis, R.J. Weld, Photosynthetic biocathode enhances the power output of a sediment-type microbial fuel cell, N. Z. J. Bot. 52 (2014) 48–59.

[63] L. Xiao, E.B. Young, J.A. Berges, Z. He, Integrated photo-bioelectrochemical system for contaminants removal and bioenergy production, Environ. Sci. Technol. 46 (2012) 11459–11466.
[64] E. Powell, R. Evitts, G. Hill, J. Bolster, A microbial fuel cell with a photosynthetic microalgae cathodic half cell coupled to a yeast anodic half cell, Energy Sources, Part A 33 (2011) 440–448.
[65] B. Logan, K. Rabaey, Conversion of wastes into bioelectricity and chemicals by using microbial electrochemical technologies (vol 337, pg 686, 2012), Science 339 (2013) 906.
[66] E. Bazdar, R. Roshandel, S. Yaghmaei, M.M. Mardanpour, The effect of different light intensities and light/dark regimes on the performance of photosynthetic microalgae microbial fuel cell, Bioresour. Technol. 261 (2018) 350–360.
[67] H.T. Tse, S. Luo, J. Li, Z. He, Coupling microbial fuel cells with a membrane photobioreactor for wastewater treatment and bioenergy production, Bioprocess Biosyst. Eng. 39 (2016) 1703–1710.
[68] B. Bordbar, A. Khosravi, R. Azin, A review on sustainable hybrid water treatment processes, in: Proceedings of the 3rd Biennial Oil, Gas, and Petrochemical Conference (OGPC2020), Bushehr, Iran, 28–30 December, 2020.
[69] E.M. Hoek, M.T.M. Pendergast, A.K. Ghosh, Nanotechnology-based membranes for water purification, in: Nanotechnology Applications for Clean Water, Elsevier, 2014, pp. 133–154.

CHAPTER 10

Algae a valuable biomass for bioethanol production

Junaid Ahmad[a], Muhammad Faisal Siddiqui[a], Lakhveer Singh[b], Farhana Maqbool[a], Sadia Qayyum[a], Ihsan Ullah[c], Fazal Adnan[d], Ziaur Rahman[e], and Muhammad Ajmal Shah[f]
[a]Department of Microbiology, Hazara University, Mansehra, Pakistan
[b]Department of Chemistry, Sardar Patel University, Mandi, Himachal Pradesh, India
[c]Department of Biological Sciences, Faculty of Science, King Abdulaziz University, Jeddah, Saudi Arabia
[d]Atta Ur Rahman School of Applied Biosciences, National University of Sciences & Technology, Pakistan
[e]Department of Microbiology, Abdul Wali Khan University Mardan, Pakistan
[f]Department of Pharmacy, Hazara University, Mansehra, Pakistan

1 Introduction

For the world's growing population, renewable energy is a major challenge. Population growth is expected to last at least several decades. The proportion of fossil fuels utilized to serve the demand for automobiles, fuels, and industries will likely grow in lockstep with energy consumption. The price of petroleum fuels has climbed owing to the daily depletion of fossil fuel supplies [1]. There are different waste, wastewater treatment technologies [2–13], and bioenergy production technologies [14–18]. Moreover, because of the increasing use of fuel supplies, different environmental problems, like climate change, have occurred. Biofuels have experienced tremendous growth due to hiking energy requirements and climate change problems. To fulfill the increasing energy requirements, eco-friendly energy resources that are incorruptible, reliable, substitutable, limitless, economical, and create low global emissions are necessary [19,20]. Renewable energy is a crucial initial stage to long-term stability. Biofuels include liquid biofuels such as bioethanol, biodiesel, and decomposition oils, along with gasses such as biogas (methane) and solids biofuels such as coal and burning woody biomass. Biomass conversion into methanol, ethanol, biodiesel, methane, and hydrogen [21].

Biofuels generated by biomass have several environmental advantages. Biofuels are necessary since Petro fuels may be depleted in the future. In many developing and industrialized countries, biofuels are crucial for reducing dependence on foreign oil, lowering greenhouse gas (GHG) emissions, and achieving regional goals. Biofuels have enhanced energy supplies, decreased market volatility, lower emissions, and eased socio-economic problems [22]. Conventional biofuels have higher expenses, which result in higher sale prices. However, government help in the financial context and income tax schemes has assisted many kinds in entering the consumption fuels sector, such as glucose biofuel in

Algae Based Bioelectrochemical Systems for Carbon Sequestration, Carbon Storage, Bioremediation and Bioproduct Generation
https://doi.org/10.1016/B978-0-323-91023-1.00012-0

Brazil. In recent decades, the production of bioethanol has grown across the globe. Similarly, production climbed from 1 billion liters in 1975 to 86 billion liters in 2010, with a goal of 1.6 trillion liters by 2020 [23]. The sustainability of biofuel has been challenged due to the scarcity of groundwater and the overuse of agricultural land. Maize, legumes, grain, barley, and sugar cane could be in limited supply due to the massive use of agricultural land for the production of bioethanol. As a reason, there are numerous discussions and debates about their long-term sustainability.

Algae are getting popular as a renewable biomass resource for the production of bioethanol to overcome these challenges [24–26]. This reveals that microalgae can thrive in a wide range of conditions. As a result, utilizing microalgae for fuel production as a sustainable energy source will ensure energy status and self. This chapter will help readers better grasp the current algal biomass bioethanol methods and how they will identify potential pollution and sustainable current problems.

2 Algae as feedstock for bioethanol production

Algae generate 5–10 times higher energy than ground crops due to their better photosynthesis rate. Lignin provides a physical barrier to the hydrolysis process if it is not pretreated. This feature of algae is beneficial throughout the pretreatment and enzymatic hydrolysis stages of the ethanol synthesis process [27]. Microalgae are grown in a controlled environment, such as an open pond or closed photobioreactor. They're also raised in a high-nutrient, high-CO_2 atmosphere. The biomass of growing algae is prepared in the same way as other lipid-based feedstocks to create biofuel. Bioethanol is produced by cells that eat carbohydrates.

2.1 Algae culture

Algae culture is the process of growing algae-like aquaculture. Algae need only sunlight, dissolved substances, and carbon dioxide to live. Axenic (bacteria-free) growth and monoalgal culture are both challenging. Unlike plants, algae spread at a breakneck speed (doubling every 24 h) [28].

2.2 Cultivation of algae

Open-pond microalgae agriculture, whether man-made or wild, is straightforward and expensive. Algal biomass can be produced in a wide range of environments, such as labs and ponds.

2.3 Laboratories for cultivation

Recent studies suggest that algal blooms of bioethanol may be grown in the laboratory in sterile environments and on a selection of culture conditions [29]. Algae can be grown in plastic containers and tubs. Oxygenation, carbon dioxide concentration, pH, temperature of the growing medium, and availability of nutrients are all influence algal biomass production in the lab. These variables are controlled to optimize biomass production. At a pH of 5.0–6.0 and a temperature of 30°C, *Saccharomyces cerevisiae* yielded the most ethanol in one experiment [30].

2.4 Open ponds

Open-pond algae cultivation, whether wild or synthetic, has been widely investigated and it is the most broadly applied described procedure [31]. In current history, many companies had increased their focus on improving the production of bioethanol. From one pond varying in size from 1 ha to many hectares, they introduce algae to natural sunlight radiation. Tanks, deep large ponds, raceway ponds, and round ponds are examples of open ponds. The elongated closed-loop circulation tubes in these systems regulate algal productiveness. However, open pond culture raises concerns about land management, water system, and climate adaptation. Significantly increasing algal biomass technology, according to one study, involves challenges such as poor cell density and high condensation costs [32]. Since the 1950s, open pond techniques have been applied to generate industrial-scale algae. This method distinguishes between natural and artificial bodies of water (lakes, lagoons, and ponds) [33]. The ecosystem is harmed by fungi, bacteria, protozoa, and microalgae. Furthermore, the system's weak agitation methods and sunlight limitations are significant drawbacks. While algae are farmed economically in industrialized countries, temperatures and sun energy levels vary throughout the year in most places. As a result, mass algal growth outside during the year is challenging. Because there are no winters or cold seasons in tropical developing countries, open cultures of algae can be studied for long periods [34] examined *Tetraselmis* sp. On the 32nd day, M8 biofuel extraction using open pond system (OPS) yields approximately 2.8×10^6 cells/mL. As stated by Ref. [35], a raceway pond uses approximately 1 W/m^3, whereas tubular photobioreactor (PBR) uses approximately 500 W/m^3.

2.5 Closed photobioreactors (PBRs)

Photobioreactors, or closed systems, are transparent containers in which microalgae are cultured [36]. Bottles or polyvinyl chloride (PVC) are frequently used for these containers. The three types of polyethylene are low-density polyethylene (LDPE), high-density polyethylene (HDPE), and PVC. The substance was using to build a PBR has an impact on its lifetime and affects its original value. Open-system reactors have major benefits over photobioreactors such as decreased chance of species contamination, increased productivity, lower harvesting costs, lower CO_2 waste, and better monitoring of growth parameters such as temperature and pH [37].

Closed photobioreactors are ideal for pure culture growth (PBRs). For the PBR to work effectively and safely, parameter optimization is necessary. In closed systems, adequate blending is usually forced to minimize biomass loss due to shadowy zones inside the reactor. Nevertheless, light penetration is sufficient in open systems and therefore can reach a depth of around 20 cm [38]. Closed systems are less likely to be contaminated because they are not exposed to the air [39]. Due to the design, these cultivation methods are normally placed vertically, horizontally, or at an angle and are operated by naturally occurring light. Closed systems can produce sensitive algae strains which can be used to give the organisms the proper environment and growth conditions. They are infrequently utilized for large-scale production since they require more development and maintenance than raceways [40,41] in subtropical Australia, researchers examined five distinct microalgae-based production environments (high-rate ponds, low-high tubes, and low and high flat panels PBRs). There are several types of bioreactors include tubing, plates, and bubbling columns, which are all used in closed bioreactors, as discussed below.

2.5.1 Tubular PBR

Tubular PBR is a liquid system comprised a net of clear or invisible tubes (with interior lights). Tubes can be straight, helical, horizontal, slanted, cylindrical, or spiral in bioreactors. The standard width of solar collectors is 0.1 m. The tubes link the gas and heat exchange systems [42]. While using outside tubular photobioreactors, heat transfer can be installed to keep temperatures low. Sunlight has to be capable of penetrating sufficiently enough into a thick growing medium for the photobioreactor to produce biomass at a high rate [43,44]. The tubular photobioreactor was found to become the most effective PBR for producing microalgae. A new culture medium is constantly added to a tube PBR, but if the same volume of microalgae biomass broth is constantly removed.

2.5.2 Flat PBR

That's how the initial fermenter appeared. The photosynthesis efficiency is highest in the zone with more light. A thin layer of material floats on a transparent layer in this circumstance. Surfaces can be vertical or sloped. These bioreactors have a restricted capacity for temperature adjustment as contrasted to tubular bioreactors [42].

2.5.3 Bubble column PBR

A bubble column photobioreactor is made up of a series of transparent cylindrical columns that are vertically aligned. The air is introduced at the bottom of the bubble column reactor. It has a 20–30 cm diameter to allow for algae growth [43].

2.5.4 Hybrid PBR

A hybrid PBR incorporates two or even more PBR kinds. A hybrid type wants to capitalize on the distinct characteristics of the constituent PBRs due to the limitations of each PBR [40].

2.6 Hybrid cultivation

Another way of producing is hybrid farming. Various improvement stages are combined into two stages in PBR and open pond systems [38].

2.7 Harvesting of algae

Algal biomass should be focused on microalgae-based harvest in favor of downstream processing to be cost-effective [45].

2.7.1 Flocculation

Flocculation is essential for obtaining the correct particle size in bulk algae harvesting. It's a step before more advanced harvesting methods like flotation, gravity sedimentation, or filtration. The greatest method for harvesting algae is flocculation, which can handle vast amounts of suspension and a wide variety of algae. Although algae species-specific flocculants are also feasible, microbes or chemicals can flocculate [28,46]. Researchers recently proved that using *Aspergillus niger* as a pelletizing agent was a promising approach for collecting microalgae easily and efficiently, getting 49% *Chlorella vulgaris*.

2.7.2 Electrochemical process

One approach is to gather microalgae using an electrochemical procedure rather than chemicals. Charged algae are ejected from the liquid by an electromagnetic current. In water, electrolysis generates hydrogen, which attaches to algal flocs and propels them forward [28].

2.7.3 Ultrasound

Despite significant influences on algal flocculation, this strategy is restricted in comparison with other methods that stimulate growth at levels up to 20 times the quantity of food [28].

2.7.4 Flotation

Flotation works by deceiving algal cells with small gas bubbles that can be eliminated without using chemicals. When an algal strain's fatty acid content rises, this will rise to the surface of the water, which can be assisted by adding air bubbles [47].

2.7.5 Filtration

Diameter or sheath size is the primary defining characteristic of membrane filtration such as microfiltration (0.1–10 m), macrofiltration (10 m), reverse osmosis (0.001 m), and ultrafiltration (0.02–0.2 m). It takes less pressure to move liquid over a larger membrane opening [1].

2.7.6 Centrifugation

Centrifugation is a simple, rapid, and ubiquitous method for collecting microalgae from liquid broth. Gravity and shearing pressures can kill microalgal cells. The time and energy necessary for large-scale harvesting make this method unsuitable [48]. Photoautotrophic, heterotrophic, photoheterotrophic, and mixotrophic microalgae are all grown differently [49].

3 Bioethanol production using algal biomass

The harvested algae biomass is processed by using dehydration and extraction processes to produce bioethanol, as discussed below.

3.1 Dehydration of algal biomass

The moisture content of the algae harvested is over 90%. As a reason, dry algae with a 50% moisture content are critical for making a firm, easy-to-handle product. Untreated algal biomass (typically 5%–15% dry dense mass) is promptly dehydrated to reduce water consumption before oil extraction. Dehydration is commonly accomplished by using an appropriate drying process. Limited shelf desalination technologies including sun drying, spray drying, freeze drying, and fluidized bed drying [50,51] are typical of algal biomass drying techniques. Each drying technique will have its own set of benefits and drawbacks. For example, sun drying algal biomass can be an expensive decalcification technology, but it has limitations such as drying durations, huge drying area needs, and losses. Spray drying is commonly utilized to remove high-value elements, even though it is expensive and results in a large loss of algal colors [52]. Crystallization may also be an effective method of dehydration since it allows for oil separation. It is, however, both expensive and hard to implement on a broad scale.

3.2 Extraction of bioethanol

Bioethanol is generally made using microalgae biomass by fermentation. Fermentation is the conversion of sugars, starches, and cellulose in biomass to ethanol [52]. The starch in the feedstock is converted to glucose, that is, then mixed with fungus and water in huge fermentation processes and maintained properly. Bioethanol can be produced when fungus gets broken into sugar [52]. Water as well as other impurities are removed first from thinned alcoholic products (10%–15% of ethanol) using a distillation technique. Pure bioethanol is evaporated and converted into a liquid that can be used in cars as a supplement or replacement for gasoline (95% volume for one purification) [53]. Ability to extract fuels from algal blooms feedstock can be accomplished in several ways. Supercritical water extraction from algal wet material is one means of obtaining algal blooms fuels. With such a water content of roughly 80%, this extract can assist algae without the need for substantial dehydration and drying. The extraction technique has also been claimed to include hexane and other organic solvents [54].

4 Bioethanol synthesis at industrial scale and its limitations

In the coming years, the international demand for algae bioethanol is expected to increase. Biofuels, petroleum, and agriculture sectors are all interested in algal biomass for biofuel production. According to sources, the world's main biofuel producers, including the United States and Europe, are still unable to reach their biofuel goals due to the absence of maize, soy, or rapeseed production. They do have the ability to produce algal biomass for long-term biofuel generation, though algae and other nonfood renewable sources will have more opportunities to reach bioethanol targets because of the long-term need for biofuels in the United States, Europe, and Asia. A variety of production issues must be resolved before algae-based fuels may be marketed on a bigger scale. One of the major barriers to algal bioethanol commercialization is the high cost of infrastructure. This massive investment is required not only to increase productivity but also to reduce the risk of microalgae contamination [55].

Other important features for the efficient and cost-effective usage of microalgae in biofuels are high lipid content, low sensitivity to contaminants, temperature resistance, and salt tolerance. The marketing of algal bioethanol to the public at large is likewise limited. One of these limits is the cost of production, which is represented in the price of biofuel. Open pond systems have much lower initial investment than photobioreactors. Photobioreactors, however, can be moved, reducing the danger of large capital expenditure. Liners account for about 24%–75% of open pond system construction costs, making them a significant cost element. To reduce processing costs, capital equipment capacity should be developed so that it is not idle for long periods. Freshwater, CO_2, and fertilizer supplies all have an impact on profitability, contributing for up to 30% of algal biofuel costs [56,57].

Nutrient input costs can be reduced by using CO_2-enriched waste streams from nearby industries or power plants, which contain enough of growing nitrogen and phosphorus. As a result, using wastewater as a nutrient source could reduce the price of algal biofuel, which is currently being researched. Furthermore, wastewater treatment credits may be redeemed if wastewater is utilized, potentially saving up to 20% on energy from other sources [58]. The algal biofuel supply chain must be efficient to increase profitability. According to one study, increasing biomass productivity by a factor of two might reduce algal biofuel production costs by 26%–32% or 40%–42%, respectively. Strain selection has an impact on profitability due to variable productivities based on lipid levels and the potential to produce coproducts. Harvesting and extraction are two major factors that influence biofuel prices [59]. A hydrotreater, a hydrogen plant, and natural gas imports would all be eliminated if hydrotreating were done at a central conventional refinery. Off-take agreements are expected to alter the price and profitability of algal bioethanol. The economics of algae biorefineries can be improved by manufacturing bioethanol from a single source [60].

However, because the bulk of algae biofuel coproducts are food and feed, market saturation may lead to decreasing commercial profitability. Furthermore, subsidies and carbon prices will have an impact on the future viability of algal biofuels. Even if technical limits are solved, practical barriers such as location, land use, and so on will take many years to overcome, and commercialization will be evolutionary rather than revolutionary. Human resourcefulness and capacity to create sustainable solutions to protect the existence of all living organisms on the planet, including humans, are put to the test by environmental concerns. To produce a diet, battle the greenhouse effect, and generate energy from nonfossil sources, there are obligations to protect fresh water and farmland lands [61]. Processes to convert biomass into affordable liquid biofuels are becoming more popular as fossil hydrocarbons become scarce and expensive. In recent years, there has been a major growth in worldwide biofuel production, which might be used to counteract the environmental effects of fossil fuels. Biofuels created from renewable resources are widely acknowledged as the most environmentally and economically feasible alternatives to petroleum-derived fuels and resources [62].

Renewable biofuels are essential for tackling energy security, environmental sustainability, and global and national climate change problems. Biofuel production from algae is a novel concept that may be assessed for sustainability and environmental protection. Algae for biofuel generation could be a promising bioresource due to their ability to consume a huge quantity of CO_2 while producing biomass. Aside from CO_2 reduction, renewable biofuels can provide energy facilities with zero or nearly low air pollution and greenhouse gas emissions [63]. Algal cells absorb nutrients such as nitrogen and phosphorus from several wastewater sources (for example, agricultural run-off, severe animal food processing, manufacturing, and community wastewater) to enable tolerable bioremediation of these wastewaters for ecological and commercial benefits. Using wastewater to grow algae for biofuels saves fresh water [64], as a result, natural water resources are conserved, wastewater bioremediation is facilitated, emissions are reduced, and financial benefits, as well as energy security, are realized. Because biofuels are "oxygenate," they contribute more oxygen to the petroleum blend, making it easier to ignite.

Bioethanol reduces the interfacial tension of gasoline with water, making it easier for the ethanol-gasoline nonaqueous phase liquid (NAPL) to pass through microscopic pore spaces and reach the water table. Biofuels have a long history of dehydrating both swelling and nonswelling clays, resulting in microfractures that improve clay penetrability [65].

Bioethanol can also be used to suspend sorbed benzene, toluene, ethylbenzene, and *p*-xylene (BTEX) and remobilize it. A unique bioethanol degradation process prevents the biodegradation of petroleum contaminants, particularly BTEX. Additional reducing conditions, such as anaerobic and even methanogenic environments, work against the natural attenuation of its biological oxygen demand (BOD). Producing biofuels and bioenergy from natural bioresources ensure energy supply independence and security. Using agricultural waste and leftovers as source materials decreases the likelihood of a food-fuel conflict while also creating biofertilizers and biopesticides [66]. Feedstuff supply rivalry with food, as well as the damage to natural environments caused by energy crop production, are all possible challenges. Biofuels improve air quality by reducing harmful air pollutants, greenhouse gas emissions, and acid rain-forming sulphur dioxide emissions, all of which contribute to ozone depletion [56]. Bioethanol is less hazardous and decomposable than petroleum fuel, and it emits fewer toxins into the atmosphere. Bioethanol is used to replace octane enhancers like methylcyclopentadienyl manganese tricarbonyl, aromatic hydrocarbons like benzene, and oxygenates like methyl tertiary butyl ether [67]. Biofuels' development and use as a fossil fuel substitute still require further inventive technical expansion to raise their viability by enhancing energy equilibrium, lowering discharges, and lowering manufacturing costs, which are actual substitutes that complete the future of the biofuels.

5 Conclusions and future perspectives

The expanded industrialization and motorization of the world have resulted in a ceaseless increase in energy mandates, and the overuse of fossil fuels has resulted in numerous environmental issues. Taking stock of our current situation, because of global warming and the depletion of fossil fuels, biofuels such as bioethanol are an excellent choice. Bioethanol production from algal biomass is also environmentally benign. For the production of renewable biofuels, algal cultivation in wastewater has several benefits. Primarily concerned with water ecosystem management by providing a low-cost, environmentally friendly, or practical alternative to standard energy systems for intensive wastewater treatment. Problems like the energy crises and environmental degradation are critical to the development of a sustainable contemporary society. Pollution needs to be addressed. Commercial algae cultivation is a well-established industry. Open-air and indoor cultivation are presently the most popular approaches. In the last 50 years, however, considerable progress has been made. Our understanding of marine life has improved. Due to engineering techniques, algal growth is increased. Bioreactors that will make it easier to commercialize novel algae in the next 10 years, and so algae products will be available. Biofuels will only be beneficial if they are grown in the proper environment. Biofuels have a promising future, but more research and development are needed. As soon as feasible, and promptly, sustainability requirements must be met for efficient bioethanol production from algal biomass.

Acknowledgments

The authors would like to express gratitude to Hazara University Mansehra for providing a platform and for the support.

References

[1] L. Brennan, P. Owende, Biofuels from microalgae—a review of technologies for production, processing, and extractions of biofuels and coproducts, Renew. Sust. Energ. Rev. 14 (2010) 557–577.
[2] M.F. Siddiqui, L. Singh, Z.A. Wahid, Treatment of Dye Wastewater for Water Reuse Using Membrane Bioreactor and Biofouling Control, in Waste Biomass Management–a Holistic Approach, Springer, 2017, pp. 121–136.
[3] M.F. Siddiqui, L. Singh, F. Maqbool, Z. Rahman, A. Rehman, F. Adnan, S. Qayyum, A. Khan, Microbial biofilm cell systems for remediation of wastewaters, in: Bioremediation and Biotechnology, Springer, 2020, pp. 293–303.
[4] M.F. Siddiqui, M. Sakinah, L. Singh, A. Zularisam, Targeting N-acyl-homoserine-lactones to mitigate membrane biofouling based on quorum sensing using a biofouling reducer, J. Biotechnol. 161 (2012) 190–197.
[5] M.F. Siddiqui, M. Sakinah, A.F. Ismail, T. Matsuura, A. Zularisam, The anti-biofouling effect of Piper betle extract against Pseudomonas aeruginosa and bacterial consortium, Desalination 288 (2012) 24–30.
[6] M.F. Siddiqui, M. Rzechowicz, H.-S. Oh, N. Saeidi, L.J. Hui, H. Winters, A.G. Fane, T.H. Chong, The efficacy of tannic acid in controlling biofouling by Pseudomonas aeruginosa is dependent on nutrient conditions and bacterial density, Int. Biodeterior. Biodegrad. 104 (2015) 74–82.
[7] M.F. Siddiqui, M. Rzechowicz, W. Harvey, A. Zularisam, G.F. Anthony, Quorum sensing based membrane biofouling control for water treatment: a review, journal of water, Process. Eng. 7 (2015) 112–122.
[8] M.F. Siddiqui, H.-S. Oh, M. Rzechowicz, H. Winters, T.H. Chong, A.G. Fane, Biofouling control potential of tannic acid, ellagic acid, and epigallocatechin against Pseudomonas aeruginosa and reverse osmosis membrane multispecies community, J. Ind. Eng. Chem. 30 (2015) 204–211.
[9] M.F. Siddiqui, S. Andleeb, N. Ali, P.B. Ghumro, S. Ahmed, Biotreatment of anthraquinone dye Drimarene blue K 2 RL, Afr. J. Environ. Sci. Technol. 4 (2010).
[10] M.F. Siddiqui, S. Andleeb, N. Ali, P.B. Ghumro, S. Ahmed, Up-flow immobilized fungal column reactor for the treatment of Anthraquinone dye Drimarene blue K sub> 2 RL, Afr. J. Biotechnol. 8 (2009).
[11] M. Siddiqui, L. Singh, A. Zularisam, M. Sakinah, Biofouling mitigation using Piper betle extract in ultrafiltration MBR, Desalin. Water Treat. 51 (2013) 6940–6951.
[12] I. Haider, I. Ullah, S.U. Khan, I. Munir, A. Mateen, A.-K. Naser, B.O. Al Johny, Y. Anwar, H.M. Al-Solami, T.M. Alahmadi, Characterization and cadmium detoxification dynamics of endophytic bacteria, isolated from rice plants, J. King Saud Univ. Sci. 34 (2022) 101992.
[13] Z.A. Bhatti, M. Syed, F. Maqbool, Y.G. Zhao, X. Ying, M.F. Siddiqui, Q. Mahmood, Potential of molasses substrate for bioelectricity production in microbial fuel cell with the help of active microbial community, Int. J. Energy Res. (2022).
[14] S. Thakur, L. Singh, Z.A. Wahid, M.F. Siddiqui, S.M. Atnaw, M.F.M. Din, Plant-driven removal of heavy metals from soil: uptake, translocation, tolerance mechanism, challenges, and future perspectives, Environ. Monit. Assess. 188 (2016) 1–11.
[15] L. Singh, Z.A. Wahid, M.F. Siddiqui, A. Ahmad, M.H.A. Rahim, M. Sakinah, Application of immobilized upflow anaerobic sludge blanket reactor using Clostridium LS2 for enhanced biohydrogen production and treatment efficiency of palm oil mill effluent, Int. J. Hydrog. Energy 38 (2013) 2221–2229.
[16] L. Singh, M.F. Siddiqui, A. Ahmad, M.H.A. Rahim, M. Sakinah, Z.A. Wahid, Biohydrogen production from palm oil mill effluent using immobilized mixed culture, J. Ind. Eng. Chem. 19 (2013) 659–664.

[17] L. Singh, M.F. Siddiqui, A. Ahmad, M.H.A. Rahim, M. Sakinah, Z.A. Wahid, Application of polyethylene glycol immobilized Clostridium sp. LS2 for continuous hydrogen production from palm oil mill effluent in upflow anaerobic sludge blanket reactor, Biochem. Eng. J. 70 (2013) 158–165.
[18] M.F. Siddiqui, Z. Ullah, L. Singh, F. Maqbool, S. Qayyum, I. Ullah, Z. Rahman, F. Adnan, Role of bioelectrochemical systems for bioremediation of wastewaters and bioenergy production, in: Delivering Low-Carbon Biofuels with Bioproduct Recovery, Elsevier, 2021, pp. 85–100.
[19] P.S. Nigam, A. Singh, Production of liquid biofuels from renewable resources, Prog. Energy Combust. Sci. 37 (2011) 52–68.
[20] F. Saïdane-Bchir, A. El Falleh, E. Ghabbarou, M. Hamdi, 3rd generation bioethanol production from microalgae isolated from slaughterhouse wastewater, Waste and Biomass Valorization 7 (2016) 1041–1046.
[21] A. Demirbas, Comparison of transesterification methods for production of biodiesel from vegetable oils and fats, Energy Convers. Manag. 49 (2008) 125–130.
[22] M. Rehan, A.-S. Nizami, O. Taylan, B.O. Al-Sasi, A. Demirbas, Determination of wax content in crude oil, Pet. Sci. Technol. 34 (2016) 799–804.
[23] Y. Khambhaty, K. Mody, M.R. Gandhi, S. Thampy, P. Maiti, H. Brahmbhatt, K. Eswaran, P.K. Ghosh, Kappaphycus alvarezii as a source of bioethanol, Bioresour. Technol. 103 (2012) 180–185.
[24] M.Y. Menetrez, An overview of algae biofuel production and potential environmental impact, Environ. Sci. Technol. 46 (2012) 7073–7085.
[25] F.W. Zechman, H. Verbruggen, F. Leliaert, M. Ashworth, M.A. Buchheim, M.W. Fawley, H. Spalding, C.M. Pueschel, J.A. Buchheim, B. Verghese, An unrecognized ancient lineage of green plants persists in deep marine waters 1, J. Phycol. 46 (2010) 1288–1295.
[26] J.M. Lopez-Bautista, F. Rindi, M.D. Guiry, Molecular systematics of the subaerial green algal order Trentepohliales: an assessment based on morphological and molecular data, Int. J. Syst. Evol. Microbiol. 56 (2006) 1709–1715.
[27] Y. Sun, J. Cheng, Hydrolysis of lignocellulosic materials for ethanol production: a review, Bioresour. Technol. 83 (2002) 1–11.
[28] R. Bibi, Z. Ahmad, M. Imran, S. Hussain, A. Ditta, S. Mahmood, A. Khalid, Algal bioethanol production technology: a trend towards sustainable development, Renew. Sust. Energ. Rev. 71 (2017) 976–985.
[29] H.-Y. Ren, B.-F. Liu, C. Ma, L. Zhao, N.-Q. Ren, A new lipid-rich microalga Scenedesmus sp. strain R-16 isolated using Nile red staining: effects of carbon and nitrogen sources and initial pH on the biomass and lipid production, Biotechnol. Biofuels 6 (2013) 1–10.
[30] M. Quayum, M. Ahmed, C. Naiyyum, Analysis of key factors affecting ethanol production by Saccharomyces cerevisiae IFST-072011, Biotechnology 11 (2012) 248–252.
[31] V. Ashokkumar, R. Rengasamy, Mass culture of Botryococcus braunii Kutz. Under open raceway pond for biofuel production, Bioresour. Technol. 104 (2012) 394–399.
[32] C.-Y. Chen, K.-L. Yeh, R. Aisyah, D.-J. Lee, J.-S. Chang, Cultivation, photobioreactor design and harvesting of microalgae for biodiesel production: a critical review, Bioresour. Technol. 102 (2011) 71–81.
[33] A. Bahadar, M.B. Khan, Progress in energy from microalgae: a review, Renew. Sust. Energ. Rev. 27 (2013) 128–148.
[34] R.R. Narala, S. Garg, K.K. Sharma, S.R. Thomas-Hall, M. Deme, Y. Li, P.M. Schenk, Comparison of microalgae cultivation in photobioreactor, open raceway pond, and a two-stage hybrid system, Front. Energy Res. 4 (2016) 29.
[35] F. Acién, E. Molina, A. Reis, G. Torzillo, G. Zittelli, C. Sepúlveda, J. Masojídek, Photobioreactors for the production of microalgae, in: Microalgae-Based Biofuels and Bioproducts, 2017, pp. 1–44.
[36] N.K. Singh, D.W. Dhar, Microalgae as second generation biofuel. A review, Agron. Sustain. Dev. 31 (2011) 605–629.
[37] I. Fernández, F. Acién, J. Fernández, J. Guzmán, J. Magán, M. Berenguel, Dynamic model of microalgal production in tubular photobioreactors, Bioresour. Technol. 126 (2012) 172–181.
[38] J.-S. Chang, P.-L. Show, T.-C. Ling, C.-Y. Chen, S.-H. Ho, C.-H. Tan, D. Nagarajan, W.-N. Phong, Photobioreactors, in: Current Developments in Biotechnology and Bioengineering, Elsevier, 2017, pp. 313–352.

[39] Q. Zhang, X. Wu, S. Xue, K. Liang, W. Cong, Study of hydrodynamic characteristics in tubular photobioreactors, Bioprocess Biosyst. Eng. 36 (2013) 143–150.
[40] M. Egbo, A. Okoani, I. Okoh, Photobioreactors for microalgae cultivation–an overview, IJSER 9 (2018) 65–74.
[41] J. Wolf, E. Stephens, S. Steinbusch, J. Yarnold, I. Ross, C. Steinweg, A. Doebbe, C. Krolovitsch, S. Müller, G. Jakob, Multifactorial comparison of photobioreactor geometries in parallel microalgae cultivations, Algal Res. 15 (2016) 187–201.
[42] R.A. Voloshin, M.V. Rodionova, S.K. Zharmukhamedov, T.N. Veziroglu, S.I. Allakhverdiev, Biofuel production from plant and algal biomass, Int. J. Hydrog. Energy 41 (2016) 17257–17273.
[43] A.R. Sirajunnisa, D. Surendhiran, Algae–a quintessential and positive resource of bioethanol production: a comprehensive review, Renew. Sust. Energ. Rev. 66 (2016) 248–267.
[44] A. Bajhaiya, S. Mandotra, M. Suseela, K. Toppo, S. Ranade, Algal biodiesel the next generation biofuel for India, Asian J. Exp. Biol. Sci 4 (2010) 728–739.
[45] M. Bilad, H.A. Arafat, I.F. Vankelecom, Membrane technology in microalgae cultivation and harvesting: a review, Biotechnol. Adv. 32 (2014) 1283–1300.
[46] S.O. Gultom, C. Zamalloa, B. Hu, Microalgae harvest through fungal pelletization—co-culture of Chlorella vulgaris and Aspergillus niger, Energies 7 (2014) 4417–4429.
[47] A. Singh, P.S. Nigam, J.D. Murphy, Renewable fuels from algae: an answer to debatable land based fuels, Bioresour. Technol. 102 (2011) 10–16.
[48] D. Surendhiran, M. Vijay, Microalgal biodiesel-a comprehensive review on the potential and alternative biofuel, Research Journal of Chemical Sciences 2231 (2012) 606X.
[49] M.R. Javed, M.J. Bilal, M.U.F. Ashraf, A. Waqar, M.A. Mehmood, M. Saeed, N. Nashat, Microalgae as a feedstock for biofuel production: Current status and future prospects, in: Top 5 Contributions Energy Research and Development, 2019, pp. 1–39.
[50] E.M. Grima, E.-H. Belarbi, F.A. Fernández, A.R. Medina, Y. Chisti, Recovery of microalgal biomass and metabolites: process options and economics, Biotechnol. Adv. 20 (2003) 491–515.
[51] J. Prakash, B. Pushparaj, P. Carlozzi, G. Torzillo, E. Montaini, R. Materassi, Microalgal biomass drying by a simple solar device*, Int. J. Sol. Energy 18 (1997) 303–311.
[52] P. McKendry, Energy production from biomass (part 2): conversion technologies, Bioresour. Technol. 83 (2002) 47–54.
[53] V. Patil, K.-Q. Tran, H.R. Giselrød, Towards sustainable production of biofuels from microalgae, Int. J. Mol. Sci. 9 (2008) 1188–1195.
[54] A. Singh, S.I. Olsen, A critical review of biochemical conversion, sustainability and life cycle assessment of algal biofuels, Appl. Energy 88 (2011) 3548–3555.
[55] J.N. Rogers, J.N. Rosenberg, B.J. Guzman, V.H. Oh, L.E. Mimbela, A. Ghassemi, M.J. Betenbaugh, G.A. Oyler, M.D. Donohue, A critical analysis of paddlewheel-driven raceway ponds for algal biofuel production at commercial scales, Algal Res. 4 (2014) 76–88.
[56] A.F. Clarens, E.P. Resurreccion, M.A. White, L.M. Colosi, Environmental life cycle comparison of algae to other bioenergy feedstocks, Environ. Sci. Technol. 44 (2010) 1813–1819.
[57] W. Zhou, P. Chen, M. Min, X. Ma, J. Wang, R. Griffith, F. Hussain, P. Peng, Q. Xie, Y. Li, Environment-enhancing algal biofuel production using wastewaters, Renew. Sust. Energ. Rev. 36 (2014) 256–269.
[58] T.J. Lundquist, I.C. Woertz, N. Quinn, J.R. Benemann, A Realistic Technology and Engineering Assessment of Algae Biofuel Production, Energy Biosciences Institute, 2010, p. 1.
[59] J.W. Richardson, M.D. Johnson, R. Lacey, J. Oyler, S. Capareda, Harvesting and extraction technology contributions to algae biofuels economic viability, Algal Res. 5 (2014) 70–78.
[60] L. Zhu, E. Hiltunen, E. Antila, J. Zhong, Z. Yuan, Z. Wang, Microalgal biofuels: flexible bioenergies for sustainable development, Renew. Sust. Energ. Rev. 30 (2014) 1035–1046.
[61] S. Prasad, A. Singh, N. Jain, H. Joshi, Ethanol production from sweet sorghum syrup for utilization as automotive fuel in India, Energy Fuel 21 (2007) 2415–2420.
[62] Y. Chisti, Biodiesel from microalgae, Biotechnol. Adv. 25 (2007) 294–306.

[63] C. Bougrier, J.-P. Delgenes, H. Carrère, Combination of thermal treatments and anaerobic digestion to reduce sewage sludge quantity and improve biogas yield, Process Saf. Environ. Prot. 84 (2006) 280–284.
[64] L. Rodolfi, G. Chini Zittelli, N. Bassi, G. Padovani, N. Biondi, G. Bonini, M.R. Tredici, Microalgae for oil: strain selection, induction of lipid synthesis and outdoor mass cultivation in a low-cost photobioreactor, Biotechnol. Bioeng. 102 (2009) 100–112.
[65] C. Ugwu, H. Aoyagi, H. Uchiyama, Photobioreactors for mass cultivation of algae, Bioresour. Technol. 99 (2008) 4021–4028.
[66] A. Singh, D. Pant, N.E. Korres, A.-S. Nizami, S. Prasad, J.D. Murphy, Key issues in life cycle assessment of ethanol production from lignocellulosic biomass: challenges and perspectives, Bioresour. Technol. 101 (2010) 5003–5012.
[67] P. Champagne, Feasibility of producing bio-ethanol from waste residues: a Canadian perspective: feasibility of producing bio-ethanol from waste residues in Canada, Resour. Conserv. Recycl. 50 (2007) 211–230.

CHAPTER 11

Evaluation and optimization of biogas production from de-oiled microalgae *Botryococcus braunii* grown in microbial fuel cell

Kowthaman Chakrapani Nagappan[a], Manaswee Suttipong[b], and Durga Madhab Mahapatra[c]
[a]Center for Waste to Wealth Management, Department of Mechanical Engineering, Centurion University of Technology and Management, Odisha, India
[b]Department of Chemical Technology, Faculty of Science, Chulalongkorn University, Bangkok, Thailand
[c]Department of Chemical Engineering, Energy Cluster, School of Advanced Engineering (SoAE), University of Petroleum and Energy Studies (UPES), Dehradun, Uttarakhand, India

1 Introduction

Fossil fuel contributes 86% of world energy demand, resulting in environmental degradation such as emitting greenhouse gases. On the other hand, several renewable resources are providing a positive effect on carbon neutrality and sustainability, but are expensive in terms of installation and operational process. In this scenario, microbial fuel cells coupled with photosynthetic algae could be developed as alternative sustainable energy systems [1,2]. Microalgae can be used as a cathode biofilm compartment in a microbial fuel cell, which provides electron acceptor oxygen for its smooth function [3]. Several studies have been carried out on such microbial fuel cells. These include algae-based single, double, and triple chambered configurations, built in a variety of shapes with Plexiglas and acrylic material to allow light penetration [4]. These fuel cells can also be used in electrochemical systems and carbon sequestration for electricity generation [5]. On the other hand anaerobic digestion is a promising technique for biogas production. Several feedstocks, such as edible, nonedible, and microorganisms, can be used for biogas production. Microalgae are microorganisms grown under photosynthetic conditions, which produce protein, lipids, and polysaccharides [6–8]. There are 300,000 species of microalgae found all over the world; among these, several species are used for waste water treatment, and as medicines, protein factors, and fuels in engines. The lipid content in the microalgae varies from species to species. The lipid extracted

Algae Based Bioelectrochemical Systems for Carbon Sequestration, Carbon Storage, Bioremediation and Bioproduct Generation
https://doi.org/10.1016/B978-0-323-91023-1.00004-1

from the microalgae is used for biodiesel production and residues are converted into useful products, namely medicinal products, fertilizers, and biogas.

Sarat et al. [9], and Chanakya et al. [10], reported that the rich content of proteins and polysaccharides in microalgae species were suitable for anaerobic biogas production processes. From the literature, Cristina et al. [11] concluded that biogas production has been carried out for several decades, but waste microalgae is less commonly used as a substrate for biogas production. This is due to its suboptimal C/N ratio, high water content, thick cell wall, and low biodegradability. The second generation feedstocks, which are nonedible crops, have not been commercially used due to deforestation and other environmental issues; whereas, the third generation feedstock derived from microorganisms, and particularly microalgae, has several advantages, such as high lipid content, greater variety of species, and can grow in brackish water. Microalgae is a promising substrate used for biofuel, particularly biodiesel and biogas production. Specifically, *Chlorella vulgaris*, *Scenedesmus*, *Chlorococcum* sp., and *B. braunii* are high in lipid content and are used for biodiesel production [12–14]. Among the three, *B. braunii* is less investigated for biogas production, especially de-oiled substrate. Which is a novel idea for biofuel applications.

The production of microalgae has also increased due to its highly photosynthetic nature, with an efficiency of 4%–5%, as compared to terrestrial crops' efficiency of 1%–2% [12,15]. Waste microalgae is the lipid extracted microalgae collected from thimbles fixed in the Soxhlet extraction apparatus. The extracted microalgae of 70% dry weight is a suitable substrate used for biogas production. However, the main drawback for microalgae is its thick cell wall, which leads to less biodegradability due to its low hydrolysis nature. Oliveira et al. [16] investigated anaerobic digestion in seaweed and reported that seaweed's microalgae has less biodegradability in polyphenols, cellulosic fibers, and lignin materials. In order to improve the biodegradability, pretreatment methods such as thermal and thermochemical, mechanical, alkali, acid, sonication, chemical evaporation, and biological pretreatments have been implemented. Among these, the thermal pretreatment method can be used to disintegrate the cells due to its high reaction efficiency and lower cost. Keymer et al. [17] reported that under high pressure thermal pretreatment, biomethane potential is doubled, whereas thermal and chemical treatment are mainly used for dewatering and partial hydrolysis of microalgae.

Tedesco et al. [18] and Kowthaman et al. [19,20] discussed breaking the crystalline structure of plant biomass, which mainly contains lignin. Lignin inhibits the decaying of cellulose and hemicellulose, which is responsible for delaying the digestion process. The mechanical technique is a process of milling the substrate to decrease the particle size, whichincreases the surface area and enhances the biogas yield by 5%–25% and decreased the digestion time by 23%–59%. On the other hand, milling requires high energy input, which isn't compensated for by extra biogas yield. Thus, a continuous increase in energy consumption hampers the commercialization of mechanical pretreatment methods.

Lee et al. [21] studied the ultrasonic pretreatment method and reported that there was an increase of 2.2-fold in biogas production when compared with the mechanical pretreatment method. This high frequency of ultrasonic waves creates the pressure difference, which results in the generation of cavitation in the cells. The generated cavitation creates turbulence and shock waves, with effects on mixing and breaking of cell walls. Thus, the cavitation increases the reaction used to break the organisms by creating the (H^+ and OH^-), which leads to the initiation of hydrolysis and subsequently volatile fatty acids [22]. The enzymatic pretreatment method uses enzymes such as esterase, protease, cellulose, and endogalacturonase for the production of biogas. Enzymatic hydrolysis enhances the breakage of cell wall, which leads to higher biogas production by 16-fold compared to other conventional process [23]. However, biogas production can be further enhanced by improving the C/N ratio, since the algal biomass has a lower carbon/nitrogen ratio.

To balance this, cornhusk, which contains 50/1 C/N, ratio can also be a suitable substrate for co-digestion. Apart from the pretreatment, co-digestion is a novel technique used to improve the biotransformation efficiency of biogas production by optimizing the nutrients and the solid waste. When compared with single substrate, co-fermentation process with different substrate enhanced the methane production [24]. The co-digestion substrates, such as oil, grease, and fat, with municipal solid waste enhanced the biomethane potential by 257%. The maximum biogas yield also depends on several variables, namely pH, C/N ratio, macro- and micronutrients, proper mixture of co-substrates, dry matter content, and toxic compounds. Thus, in this research work, de-oiled *B. braunii* was investigated for its potential in developing microbial fuel cells and the harvested algae was used for biogas production with cornhusk as a co-digestion substrate. Furthermore, a central composite rotatable design was implemented to identify the best pretreatment parameters. The design interaction between the pretreatment temperature and treatment time was extensively studied and the theoretical values were compared with the experimental results. The physiochemical characterization of raw biomass and de-oiled biomass were examined. The produced biogas was sampled and the composition of gas determined.

2 Materials and methods

The microbial fuel cell consists of an anode and cathode chambers partitioned by a cation exchange membrane; carbon fiber veil was used as electrode material for both anode and cathode, with a surface area of 280 cm^2. The cathode electrode was fixed with the biofilm and connected to the photo bioreactor. A pump was used to circulate the algae water from the photo bioreactor to the microbial fuel cell and from the microbial fuel cell to the photo bioreactor. The growth of the algae was monitored

Fig. 1 Waste to substrate preparation.

periodically and, once the algae growth stopped, it was harvested and the lipids extracted for further processing. Nowadays, machine learning and artificial intelligence can be incorporated in bio electrochemical systems in the design of the reactor [25,26]. Hydrogen production was also investigated by using various cathodic and anodic electrodes [27]. In this experiment, the Botryococcus algal strain water is fed into the photo bioreactor and the harvested algae used to extract biofuels. Fig. 1 shows the layout of the microbial fuel cell setup.

100 g of *botryococcus* is powdered and dropped into a thimble and placed inside the reflex container. The reflex container is placed over the three-neck flask and an airtight seal is maintained to prevent the evaporation of hexane in the flask. Hexane has a boiling point of 69°C, meaning it evaporates and passes towards the reflux container to break the cell wall of the substrate present inside the thimble and is drawn back to the flask with the extracted lipids. The lipids are filtered and used for biofuel production, and the remaining biomass is washed and sundried for the further utilization, such as biogas and byproducts production. Corn is commercially available for consumers and is rich in fiber content; however, after use the cornhusks are thrown away, which produce environmental effects. In order to reduce these effects, the cornhusk can be utilized for biofuel production. In this research work, cornhusk was collected from a campus dumping yard, and was cleaned and chopped into fine pieces, as shown in Fig. 2. Mainly, cornhusk is used as a co-digestant to balance the carbon/nitrogen ratio of the digester. To start the reaction, 1 g of cow dung is used as an inoculum and dropped into the digester.

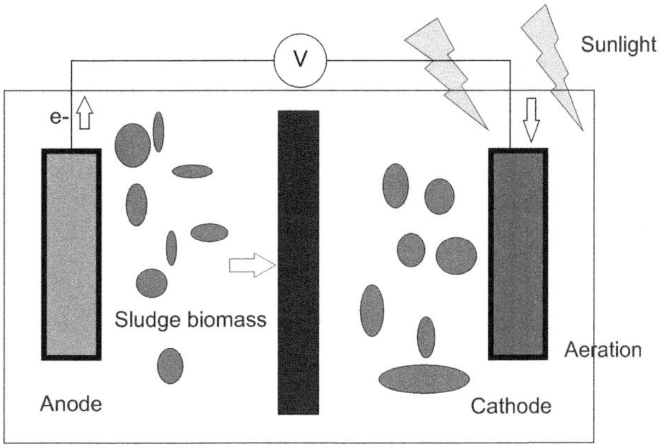

Fig. 2 Cathode reaction of algal-based microbial fuel cell.

2.1 Anaerobic biodegradability assays

An anaerobic biodegradability bath digestion setup was used to assess the biogas production with the co-digestion of cornhusk. Different parameters were considered for the optimization of the digestion process using design of experiment by response surface methodology. Waste biomass obtained from the Enviro Biotech farm, Coimbatore, is cleaned and sun dried until the substrate is free from moisture. A number of proximate and ultimate analyses were carried out to study primary information about the substrate, as shown in Table.1. Total solids were determined by drying the biomass in a furnace overnight at 105°C and the sample was weighed. For volatile solids, the sample is kept at 550°C for 15 min in a muffle furnace and placed in the silica desiccator. To determine the volatile solids, the sample is weighed before and after heat treatment in muffle furnace. This procedure was followed in the work carried out by Anil et al. [28].

2.1.1 Experimental design

All experiments were conducted in 500 mL serum bottles closed tightly with butyl rubber corks. To maintain the serum bottles in anaerobic condition, the cork is covered with an aluminum cap. A pretreated quantity of 10 gVS of waste de-oiled algae was filled in the serum bottles; to balance the C/N ratio 5 gVS chopped of corn husk is also added to the

Table 1 Characterization of various substrates for biogas production.

Parameters	Waste microalgae (%)	Cornhusk (%)	Cow dung (%)
Total solids(TS)	88	90	22
Volatile solids(VS)	62	69	70
C/N ratio	4.5/1	50/1	25/1

Fig. 3 Biogas laboratory scale setup.

serum bottles. The serum bottle is divided into 300 mL of working volume and the remaining 200 mL was maintained for occupying the biogas. To initiate the process, the acclimatized 10gVS cow dung was introduced as an inoculum in the serum bottles. Before crimping the aluminum stopper in the bottles, the bottles are kept in a water bath to maintain the digestants at thermophilic conditions of 55°C. The required amount of water was added to the serum bottle to maintain the digester at 10% total solids. Also, by adding sodium hydroxide solution, the initial pH was kept at 7.2. The substrate was agitated daily in the bottle to prevent settling.

The maximum temperature maintained for pretreatment was 110°C because, as per the work reported by Fabiana et al., thermal pretreatment temperature below 100–120°C under atmospheric conditions was shown to increase the yield. The quantification of biogas was done with the help of 20 mL bags, by replacing the bags daily with fresh ones the contamination was avoided, as butyl rubber cork resist and are crimped with the aluminum cork, as shown in Fig. 3.

2.1.2 Selection of optimization parameters

The pretreatment temperature and time were taken as primary factors to study the effect of biogas production. A design of experiments based on response surface method was implemented to optimize the process parameters. The design incorporated 13 experiments, as shown in Table 2. All the experiments were carried out in triplicate and the final results were averaged. The analysis of variance predicts the degree of freedom, sum of squares, mean squares, F-test and P-test. Also, it plots and predicts the regression coefficient equations. The polynomial equation is shown below:

$$Y = \beta_0 + \sum \beta_i Xi + \sum \beta_{ii} Xi^2 + \sum \beta_{ij} XiXj + \ldots \quad (1)$$

where Y is the overall response of the experiments predicted, X_i and X_j are the different factors involved in this experiments, β_0 is the independent coefficient term, and β_{ii} and β_{ij} represent the quadratic and interaction coefficient, respectively.

The full experiment plan and the actual yield and predicted yield are listed in Table 2.

Table 2 Different run using response surface method.

Run	A:temperature	B:time	Biogas yield actual (mL)	Biogas yield predicted (mL)
1	110	12	142	88
2	70	12	624	626
3	70	24	546	518
4	50	18	680	678
5	70	12	596	596
6	90	6	518	568
7	70	12	635	626
8	50	6	349	317
9	90	18	226	306
10	30	12	180	208
11	70	1	446	449
12	70	12	636	626
13	70	12	689	626

3 Results and discussion

A microbial system using *B. braunii* for the production of biogas requires various steps, which include the harvesting of biomass from the cathode system, and separation of algal biomass using a filtration method [7,12]. The obtained biomass is used to extract the lipid and it is characterized for biofuel applications. The de-oiled biomass was used as a substrate for biogas production. Eq. (2) Shows the developed model by design of experiments for predicting biogas production.

$$\text{Biogas yeild} = 627.93 - 43.06 \times A + 17.44 \times B - 149.26 \times AB - 119.33 \times A^2 - 35.97 \times B^2 \quad (2)$$

The results of this analysis are shown in Table 3. The biogas yield output of the response surface methodology indicated that the model fitted as a quadratic equation is significant for explaining the biogas production potential. The modeled R^2, adjusted R^2, and adequate precision are 0.9575, 0.9271, and 16.432, respectively. A precision greater than 4 indicates the model have been estimated significantly [18,29,30]. The correlation coefficient (R^2) of second order polynomial equation was determined as 0.9575. The *P*-value is the probability value of the individual and interaction factors. The *P*-value less than 0.005 indicates that the developed factor is feasible to predict the future experiments. In this experiment, pretreatment temperature, pretreatment time, and combinational effects showed positive response, which has a probability value less than 0.005. The modeled R^2, which is the coefficient of regression, showed the best interactions and individual effects of 96.75%. Thus, the developed model has best fit solution for the quadratic models. The remaining 3.25% showed that the developed

Table 3 Analysis of variance table.

Source	Sum of squares	DOF	Mean square	Stand. error	F-test	*P*-value Prob > *F*	Effects
Model	4.395E+005	5	87,905.93	22.13	31.52	0.0001	Significant
A-Temp.	21,798.41	1	21,798.41	15.40	7.81	0.0067	
B-Time	3762.28	1	3762.28	15.02	1.35	0.0036	
AB	97,032.25	1	97,032.25	25.31	34.79	0.0006	
A^2	3.267E+005	1	3.267E+005	11.03	117.13	< 0.0001	
B^2	29,816.67	1	29,816.67	11.00	10.69	0.0137	
Residual	19,525.29	7	2789.33				
Lack of fit	14,971.29	3	4990.43		4.38	0.0937	Not significant
Pure error	4554.00	4	1138.50				
Cor. total	4.591E+005	12					

or input parameters are not feasible for this experiment. The adjusted R^2 value of 0.947 showed the high significance of the model. The overall predicted residual sum and the *F*-value was found to be 31.52.

The experiment continued until there was no longer any biogas being produced at 50°C in a thermophilic environment. A 2N NaOH solution was mixed in the reactor to raise the mixture pH to 7.2. For evenly distribution of the substrate, the bottles were shaken each day. The quantity of trials performed to optimize the input variables and its actual biogas output are listed in Table 2. Fig 4. represents the cumulative biogas production for the duration of 8 days. The maximum yield for Day 1 is achieved in Run 4, in which the substrate is pretreated at 50°C for a time of 18 h; whereas minimum yield on Day 1 is Run 1 with 110°C in 12 h. Run 13 yields the highest biogas production of 689 mL/gVS, compared to the 142 mL/g VS in Run 1 of 110°C in 12 h.

Fig. 5 shows the runs compared with the biogas production. Run 13 represents the highest yield of 689 mL/gVS of biogas production because of its pretreatment temperature of 70°C, which cracks the cell wall more gently than the lowest (Run 1). Run 1 is operated at 110°C with 12 h span, and cracks the cell and spoils the fat content, thus showing lower biogas production of 110 mL/g VS. Run 1 shows the lowest biogas production of 110 mL/g VS, as compared to freshwater microalgae biomass optimized in proper temperature and inoculum ratio of *Microcystis* sp. yield of 153 mL/g VS [31].

Fig. 6 plots the residual vs predicted. The residuals are scattered randomly in a constant range of intervals across the graph. The residuals are randomly distributed over the graph at regular intervals. Since the residuals are small but have a tendency to converge towards the diagonal line, Fig. 7 shows the predicted vs. actual plot, which indicates a higher degree of similarity between the experiment's result and the actual value. The

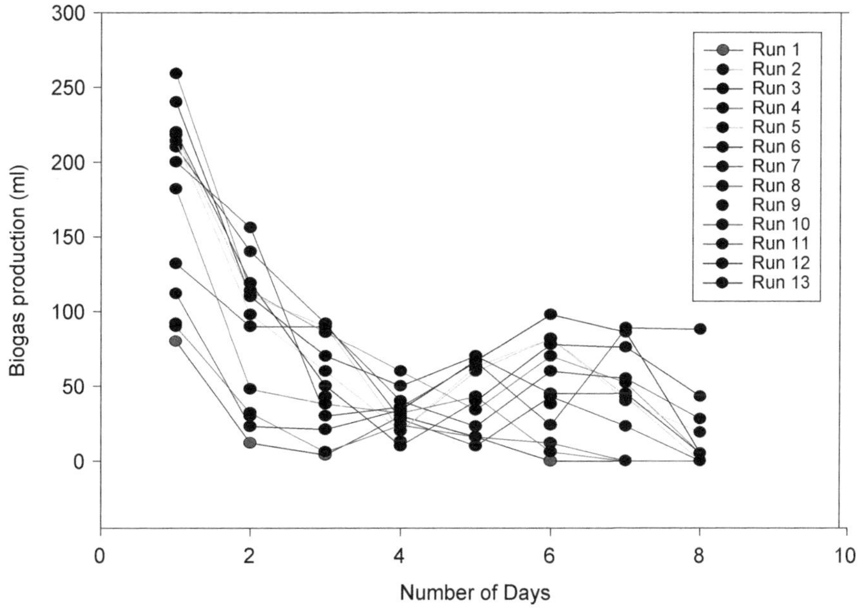

Fig. 4 Cumulative biogas production.

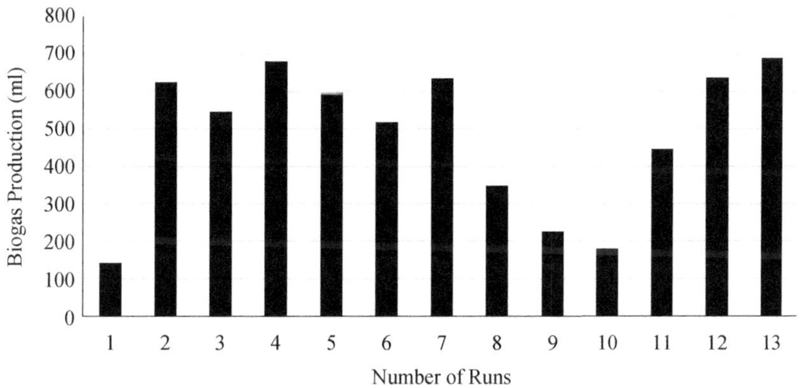

Fig. 5 Biogas yield for different trials.

model was fitted in a quadratic equation and satisfied the assumption stated by the ANOVA, leading to the conclusion that optimizing RSM is the optimum strategy for maximizing biogas output.

Fig. 8 shows the contour plots, which shows that the region between the temperatures of 60°C and 80°C gives the greater yield. Among these, 78°C and 18h shows the best predicted region for biogas production. The temperature region between 50°C and 65°C shows that there was no significant effect on increasing the yield, because maximum

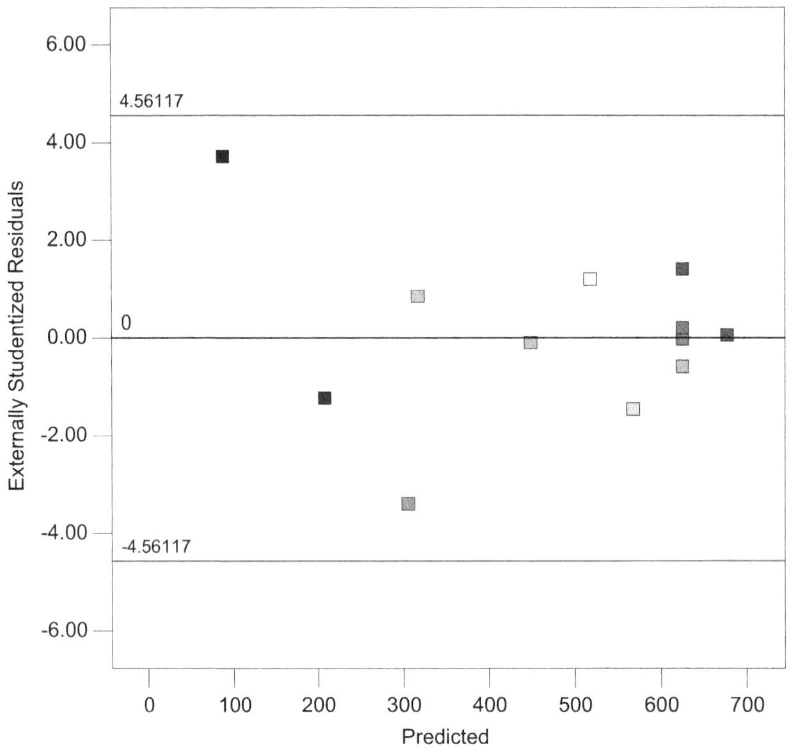

Fig. 6 Residuals vs predicted.

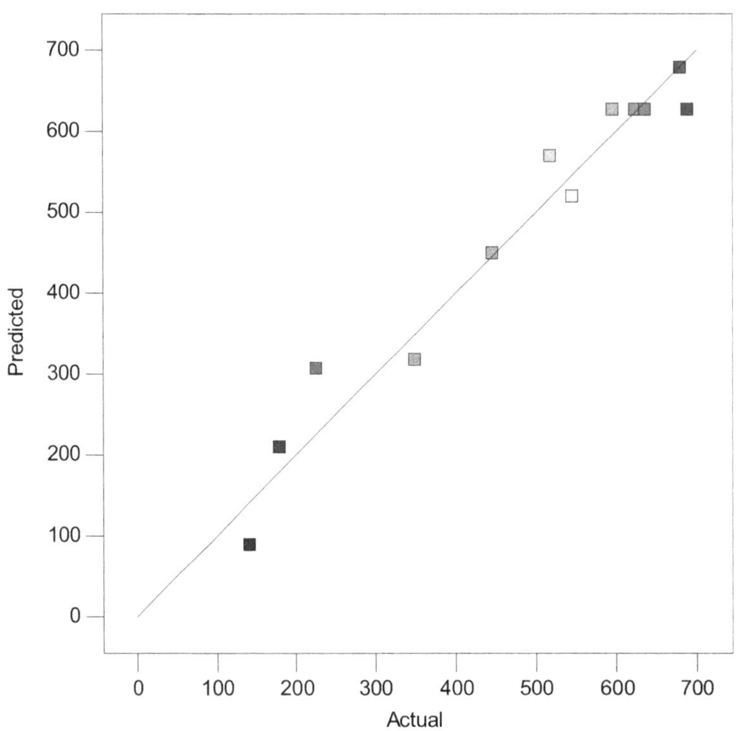

Fig. 7 Actual vs predicted.

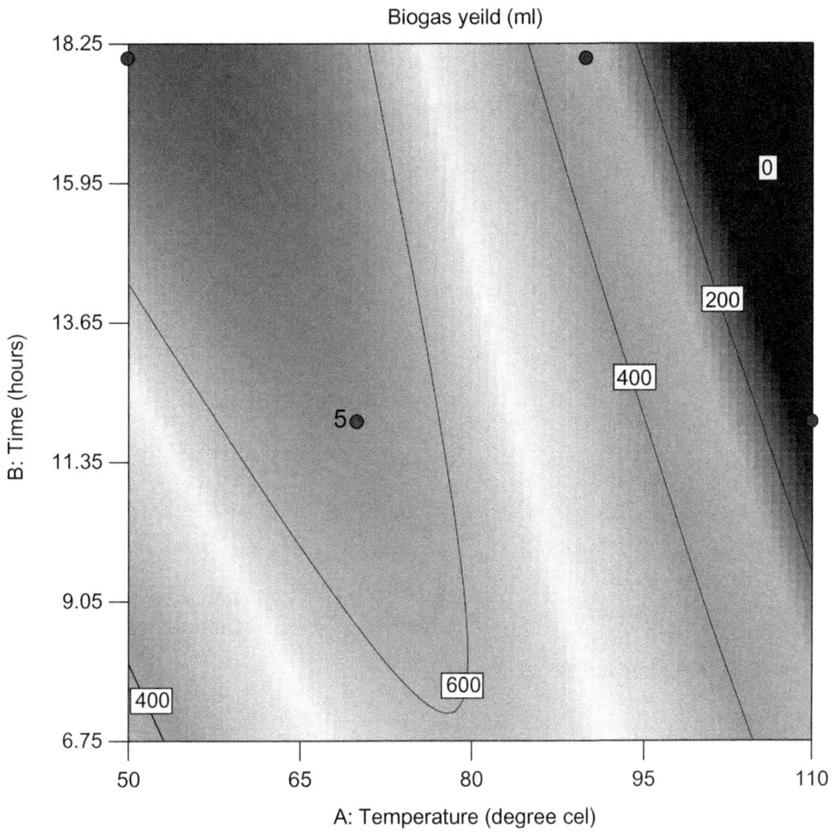

Fig. 8 Contour plots for biogas yield.

yield in this region was 400 mL/g VS, which is comparatively higher than the region between 80°C and 95°C of 200 mL/g VS. The synergetic effect of individual variables such as temperature and pretreating hours are determined for biogas output. The 3D plot shown in Fig. 9 represents the biogas yield with change in temperature and time. The surface response plot shows the peak point above the 600 mL at a temperature of 70°C and the temperature at 12°C shows the optimum region from the actual value.

3.1 Future scope

it is necessary to discuss increasing the efficiency of algae separation from the cathode compartment, as some of the algae biomass is attached to the surface of the electrode. Myriads of algal species found, extends the scope of the present work on photosynthetic growth studies, separation, filtration, extraction and other applications. Also the electrode

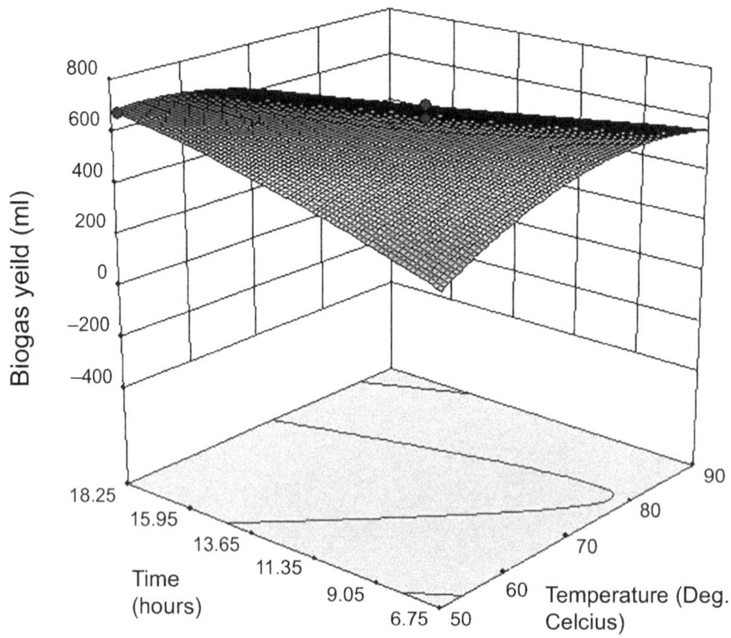

Fig. 9 Response surface plots for biogas production.

material can be studied for its morphological structure, so that the life of the electrode is predicted with respect to the different species and electrolyte used.

4 Conclusion

The present study focuses on obtaining *Botryococcus braunii* from the cathode compartment of a microbial fuel cell system, and optimizing the process parameters, such as pretreatment temperature and treatment time, for maximum biogas production. Optimization was done to save the laboratory chemical cost, for which the total number of runs was reduced. Response surface methodology best determines the optimum process parameters. Hence, the substrate contains hemicellulose, which is resistant to degrading for digestion. So, pretreating the substrate at 70°C gives maximum biogas yield of 679 mL/g VS, as compared to 110°C giving 130 mL/g VS. The maximum biogas was due to the maximum hydrolysis obtained at a temperature of 70°C. The minimum yield was achieved because of loss of mass in lipids and substrate. Hence, the microalgae biogas production achieved better yield if the C/N ratio was balanced.

References

[1] F.A.A. Alatraktchi, Y. Zhang, I. Angelidaki, Nanomodification of the electrodes in microbial fuel cell: impact of nanoparticle density on electricity production and microbial community, Appl. Energy 116 (2014) 216–222.

[2] H.S. Lee, P. Parameswaran, A. Kato-Marcus, C.I. Torres, B.E. Rittmann, Evaluation of energy-conversion efficiencies in microbial fuel cells (MFCs) utilizing fermentable and non-fermentable substrates, Water Res. 42 (6–7) (2008) 1501–1510.

[3] I. Gajda, J. Greenman, C. Melhuish, I. Ieropoulos, Self-sustainable electricity production from algae grown in a microbial fuel cell system, Biomass Bioenergy 82 (2015) 87–93.

[4] Z. Ge, L. Wu, F. Zhang, Z. He, Energy extraction from a large-scale microbial fuel cell system treating municipal wastewater, J. Power Sources 297 (2015) 260–264.

[5] Z. Chi, J.V. O'Fallon, S. Chen, Bicarbonate produced from carbon capture for algae culture, Trends Biotechnol. 29 (11) (2011) 537–541.

[6] H.N. Chanakya, D.M. Mahapatra, R. Sarada, R. Abitha, Algal biofuel production & mitigation potential in India, Mitig. Adapt. Strateg. Glob. Chang. 18 (2013) 113–136.

[7] D.M. Mahapatra, G.S. Murthy, Long term evaluation of a pilot scale multimodal algal bioprocess for treatment of municipal wastewater, J. Clean. Prod. 311 (2021) 127690.

[8] D.M. Mahapatra, H.N. Chanakya, T.V. Ramachandra, Euglena sp. as a suitable source of lipids for potential use as biofuel and sustainable wastewater treatment, J. Appl. Phycol. 25 (2013) 855–865.

[9] T. Sarat Chandra, G. Suvidha, S. Mukherji, V.S. Chauhan, S. Vidyashankar, K. Krishnamurthi, R. Sarada, S.N. Mudliar, Statistical optimization of thermal pretreatment conditions for enhanced biomethane production from defatted algal biomass, Bioresour. Technol. 162 (2014) 157.

[10] H.N. Chanakya, D.M. Mahapatra, R. Sarada, V.S. Chauhan, R. Abitha, Sustainability of large-scale algal biofuel production in India, J. Indian Inst. Sci. 92 (1) (2012) 63–98.

[11] C. Gonzalez-Fernandez, B. Molinuevo-salces, M.C. Garcia-Gonzalez, Evaluation of anaerobic codigestion of microalgal biomass and swine manure via response surface methodology, Appl. Energy 88 (2010) 3448.

[12] D.M. Mahapatra, H.N. Chanakya, T.V. Ramachandra, Treatment efficacy of algae based sewage treatment plants, Environ. Monit. Assess. 185 (2013) 7145–7164.

[13] D.M. Mahapatra, T.V. Ramachandra, Algal biofuel: bountiful lipid from Chlorococcum sp. proliferating in municipal wastewater, Curr. Sci. 105 (2013) 47–55.

[14] T.V. Ramachandra, D.M. Mahapatra, B. Karthick, R. Gordon, Milking diatoms for sustainable energy: biochemical engineering vs. gasoline secreting diatom solar panels, Ind. Eng. Chem. Res. 48, 19 (2009) 8769–8788.

[15] D.M. Mahapatra, H.N. Chanakya, T.V. Ramachandra, Bioremediation and lipid synthesis of myxotrophic algal consortia in municipal wastewater, Bioresour. Technol. 168 (2014) 142–150.

[16] J.V. Oliveira, M.M. Alves, J.C. Costa, Optimization of biogas production from sargassum sp. using a design of experiments to assess the co-digestion with glycerol and waste frying oil, Bioresour. Technol. 175 (2014) 480.

[17] P. Keymer, I. Ruffell, S. Pratt, P. Lant, High pressure thermal hydrolysis as pre-treatment to increase the methane yield during anaerobic digestion of microalgae, Bioresour. Technol. 131 (2012) 128.

[18] S. Tedesco, K.Y. Benyounis, A.G. Olabi, Mechanical pretreatment effects on macroalgae-derived biogas production in co-digestion with sludge in Ireland, Energy 61 (2013) 27.

[19] C.N. Kowthaman, P. Senthil Kumar, V.A. Mozhi, D.G. Selvan, A comprehensive insight from microalgae production process to characterization of biofuel for the sustainable energy, Fuel 310 (Part B) (2022) 122320.

[20] C.N. Kowthaman, V.A.M. Selvan, P.S. Kumar, Optimization strategies of alkaline thermo-chemical pretreatment for the enhancement of biogas production from de-oiled algae, Fuel 303 (2021) 121242.

[21] K. Lee, P. Chantrasakdakul, D. Kim, M. Kong, K.Y. Park, Ultrasound pretreatment of filamentous algal biomass for enhanced biogas production, Waste Manag. 34 (2013) 1035.

[22] C. Rodriguez, A. Alaswad, J. Mooney, T. Prescott, A.G. Olabi, Pretreatment techniques used for anaerobic digestion of algae, Fuel Process. Technol. (2015), https://doi.org/10.1016/j.fuproc.2015.06.027.

[23] T.V. Ramachandra, D.M. Mahapatra, S. Samantray, N.V. Joshi, Biofuel from urban wastewater: scope and challenges, Renew. Sust. Energ. Rev. 21 (2013) 767–777.

[24] X. Wang, G. Yang, Y. Feng, G. Ren, X. Han, Optimization feeding composition and carbon-nitrogen ratios for improved methane yield during anaerobic co-digestion of dairy, chicken manure and wheat straw, Bioresour. Technol. 120 (2012) 78.

[25] D.M. Mahapatra, P. Mishra, S. Thakur, L. Singh, Leveraging artificial intelligence in bioelectrochemical systems, Trends Biotechnol. 40 (5) (2022) 535–538.

[26] M. Karthikeyan, D.M. Mahapatra, A.S.A. Razak, A.A. Abahussain, B. Ethiraj, L. Singh, Machine learning aided synthesis and screening of HER catalyst: present developments and prospects, Catal. Rev. (2022) 1–31.

[27] P. Mishra, P. Sudarsanam, D.M. Mahapatra, A. Elmekawy, D. Pant, L. Singh, Progressions in cathodic catalysts for oxygen reduction and hydrogen evolution in bioelectrochemical systems: molybdenum as the next-generation catalyst, Catal. Rev. (2021) 1–93.

[28] A.K. Mathew, I. Bhui, S.N. Banerjee, R. Goswami, A.K. Chakraborty, A. Shome, S. Balachandran, S. Chaudhury, Biogas production from locally available aquatic weeds of Santiniketan through anaerobic digestion, Clean Technol. Environ. Policy 17 (2014) 1681.

[29] D. Mukhopadhyay, J.P. Sarkar, S. Dutta, Optimization of process parameters for the economical generation of biogas from raw vegetable waste under positive influence of plastic materials using response surface methodology, J. Biochem. Technol. 4 (2012) 549.

[30] F. Passos, E. Uggetti, H. Carrere, I. Ferrer, Pretreatment of microalgae to improve biogas production: a review, Bioresour. Technol. 172 (2014) 403.

[31] A.J. Ward, D.M. Lewis, F.B. Green, Anaerobic digestion of algae biomass: a review, Algal Res. 5 (2014) 204.

CHAPTER 12

Algae-based bioelectrochemical systems for bioremediation and co-generation of value-added chemicals

Swayam Siddha[a] and Manoj Kumar[b]
[a]School of Environment and Sustainable Development, Central University of Gujarat, Gandhinagar, Gujrat, India
[b]Department of Hydro and Renewable Energy, Indian Institute of Technology Roorkee, Roorkee, Uttarakhand, India

1 Introduction

Water contamination from the release of numerous lifestyle chemicals from municipal wastewater has long been a source of worry. Because, it has an immediate impact on human well-being and jeopardizes aquatic ecosystems especially drinking water resources [1]. The production of wine, tanned leather, textiles, and coffee in isolated rural locations is one example of a small industry or industrial unit that supports local economic growth, this discharge frequently contains high quantities of organic matter, and nutrients, in addition to metallic elements [2]. The effluents from industries, which include dyes, surfactants, minerals, and certain metals, constitute a serious threat to aquatic ecosystems and animals [3]. Surface water, soil, and groundwater are all severely contaminated by the waste or byproducts associated with many sorts of contaminants. Water that has been contaminated by industrial discharge includes harmful chemicals and other materials, rendering it unfit for usage and other purposes [4].

Research is being done on a very promising variety of microbial fuel cells (MFCs), including the incorporation of MFC technology for environmental cleaning into natural ecosystems. Recent research shows that MFCs can offer potential for biodegradable compound-based sustainable power generation [5]. A cutting-edge and practical wastewater treatment method are the utilization of bioelectrochemical systems (BESs) in the production of energy and hydrogen [6,7]. An MFC system and an electrolysis system have both been shown to significantly increase nitrogen removal from wastewater [6,8]. A particularly promising class of MFCs that involve incorporating MFC technology into ecosystems or a low-cost environmental remediation technique is now being researched [9]. Recently, there has been an increase in interest in waste energy recovery. MFCs are a potential new development in the field of sustainable energy generation [10–21]. MFCs are bio-electrochemical systems that allow for the direct conversion of chemical energy found in bio-convertible waste to electrical energy through a sequence

Algae Based Bioelectrochemical Systems for Carbon Sequestration, Carbon Storage, Bioremediation and Bioproduct Generation
https://doi.org/10.1016/B978-0-323-91023-1.00007-7

of bio-electrochemical processes in a specific fuel cell. By reducing both the energy input and the excessive sludge generation, MFC makes it possible to recover energy from the waste while treating it [19]. However, the majority of the previous investigations on MFCs used pure cultures and straightforward specified substrates. Recently, interest in utilizing wastewater such as food waste has increased [22], electroplating [12], starch processing [17], brewery [22–25], paper industry [26], palm oil mill [27], chocolate industry [28], domestic sewage [29], cellulosic waste [30,31], vegetable waste [19], wastewater from composite chemicals [20], swine waste [32], etc., as substrates.

The vast range of possible uses for microalgal technology in food, feed, fuel, pigments, and other byproducts has attracted a lot of scientific attention [33]. The adaptability of microalgae to thrive in situations with wastewater is another noteworthy characteristic [34,35]. Mixotrophy, a technique used by microalgae, allows them to take up dissolved organic carbon from wastewater while still performing photosynthesis [36]. This capability expands the range of applications for microalgae in bioremediation and CO_2 sequestration. Under mixotrophic culture, several species of *Chlorella, Scenedesmus, Nannochloropsis*, etc., have demonstrated increased harvests of biomass [37,38]. Despite the advantages, low cell strength and scarcity of efficient biomass collecting techniques remain the main barriers to the use of microalgae for generating energy.

Recent years have seen an increase in the popularity of MFCs as another green technology. By using microorganisms to catalyze biocatalytic processes, organic matter may be directly transformed into energy in MFCs [39,40]. Utilizing wastewater in MFCs provides the two-fold advantages of generating electricity and treating effluent. Additionally, it lowers the energy consumption that would otherwise be required for treating wastewater through an activated sludge process. The oxygen reduction process at the cathode is one of the main issues that restrict MFC performance [41]. Additionally, more energy is required to pump O_2 to the cathode chamber, raising the process' total costs [42]. As a result, efforts are being undertaken to use biocathodes instead of the traditional abiotic cathodes in MFCs [43–45]. In light of the aforementioned scrutiny, coupling the advantages of microalgal and MFC expertise into a sole structure acknowledged as "microbial carbon capture (MCC)" cells seem to be an appealing way to treat wastewater, sequester carbon, and produce energy in the form of bioelectricity simultaneously [46,47]. MCCs are modified microbial fuel cells that produce oxygen at the cathode using photosynthetic microorganisms [48,49]. It was previously shown that by employing microalgal biocathodes, CO_2 produced at the anode could be efficiently sequestered [47,50].

Microalgae can take the role of expensive and energy-intensive aeration systems as a root of oxygen [51], and MFC becomes an MCC as a result. *Chlorella sorokiniana* is well known for its high photosynthetic activity and lipid content, making it an excellent feedstock for the production of biodiesel and MCC applications as well [52].

2 Bioelectrochemical system

Based on the biocatalyst, BESs are classified as MFCs or enzymatic fuel cells (EFCs) [53]. Largely dependent on their use, BESs are subsequently classed as MFCs, microbial electrolysis cells (MECs), microbial desalination cells (MDCs), as well as microbial solar cells (MSCs). The MSC concept has been thoroughly discussed [54,55]. Moreover, the idea of using BESs for concurrent desalination with energy/hydrogen recovery in the form of MDC was introduced [56], which was then elaborated on by additional scientists [57,58].

BESs can turn the chemical energy of biowaste, like low-strength wastewaters and lignocellulosic biomass, into power generation or hydrogen/chemical merchandise in MFCs or alternative products generated at the cathode by an electrochemical reduction method, or MECs [59]. Electrochemical energy creation has been carefully studied as an alternative energy/power source as long as this energy demand is meant to be more sustainable and eco-friendly [60]. By producing power from "negative-value" waste streams, it is conceivable to meet the world's electricity needs, reduce pollution, and reduce the price of water and wastewater treatment [59].

3 Algae-based bioelectrochemical systems for wastewater treatment

Through the process of photosynthesis used by algae for biosynthesis and gaining energy, light is transformed into chemical energy through the processes of CO_2 reduction and H_2O oxidation. However, bioenergy is created by using the metabolic processes of bacteria to transform chemical energy into electricity. Algae based electrochemical systems take advantage of the ability of microalgal photosynthetic activity for oxygen synthesis in the cathode and/or a bioelectricity generator in the anode. As an outcome, the oxidation of the biosynthesized degradable organic molecules occurs in the anode during illumination, while in the cathode, it serves as a terminal electron acceptor by supplying bio-oxygen. Consequently, an algae-based sustainable bioenergy production system was constructed. Microalgae have enacted a vital part in the development of biosensing technology in addition to its use in microbial fuel cells. Algae-based biosensors have shown the capacity to sensitively, persistently, and multiplexedly detect the analytes of agricultural environmental and security interests [61]. *Scenedesmus dimorphus*, mixotrophic algae, has a superior ability to remove chemical oxygen demand (COD) than other MFCs because it combines an autotrophic and a heterotrophic mode of feeding [62]. Huarachi-Olivera et al. [63] discovered the importance of increased oxygen reduction reaction (ORR) activity of biocathodic *Chlorella vulgaris* utilized in A-MFC to accomplish a COD removal of 78.6% in chocolate manufacturing organic wastewater. In their study, Colombo et al. [64] showed that air-breathing-microbial fuel cells (AMFCs) using *Spirulina* biocathodes removed soluble-COD (sCOD) (effluent concentration of $359 \pm 13\,mg\,L^{-1}$) by a factor of 2 more effectively than conventional MFCs. Li et al. [65] discovered that the A-MFC had enhanced tryptophan and tyrosine degradation. With an algal biomass production of $0.56\,kg\,m^3$, Zhang et al. [66] achieved an 87.6% nitrogen removal efficiency in a sediment-type photomicrobial algal fuel cell. Using the algae *Golenkinia* sp. SDEC-16, Hou et al. [67] removed almost 90% of TN from anaerobically digested effluent from kitchen waste (ADEKW).

3.1 Microalgae microbial fuel cell system

MFCs are a cutting-edge and revolutionary innovation that create power from waste and renewable biomass at the anode end via microbial catalytic reactions [54,68]. It also aids in the bioremediation of some pollutants and nutrients in wastewater [69]. It may also be used to reclaim heavy metal ions, decolorize dyes, and produce bioenergy like biomethane, biohydrogen, and even biomass [45,69]. As a result, it provides both power generation and wastewater treatment, making the process more ecologically and financially sound as a whole [70]. The reduction of oxygen to create water can be catalyzed by abiotic cathode processes. However, in such processes, the introduction of price components, such as platinum, lessens the process's economic unviability [54]. The potential of microalgae in various MFC systems for electricity generation is being studied extensively. Bajracharya et al. [71] have offered outstanding assessments of the numerous MFC variants. Experts have also discussed the fundamental distinctive characteristics of every system. Algae in MFC technologies are useful since these can also be used as efficient electron acceptors during photosynthetic processes at the cathodic end of the cell or as an electron donors at the anodic end of the cell. They can also eliminate organic material [72]. Algae-based MFCs are the result of a syntrophic interaction between bacterial populations and algal biomass, and this system uses very little net energy. The process of algal MFC comprises oxidizing biodegradable materials, producing electrons at the anode, and emitting CO_2 at the cathode [73,74].

3.2 Single-chamber algae-microbial fuel cell system

Single-chambered microbial fuel cells (SCMFCs, Fig. 1) typically have a cubic shape, although they can also occasionally have a cylindrical shape [67,75]. Only one chamber (membrane-less) MFCs have received much prior research. Without the need for "electron shuttle mediators," photosynthetic biocatalysts have demonstrated the capacity to move electrons to the electrode surface [76]. Without employing membranes, the blue-green microalga *Spirulina platensis* has been examined. This MFC system generated electricity with an output of $1.64\,mW\,m^2$ in the absence of light and $0.132\,mW\,m^2$ in the presence of light [77]. As a result, more electric power is created in the dark than in the light. Because of these characteristics, particularly better power yields in high light conditions, single-chambered MFC can function for prolonged periods. This specific MFC design is ideal for increasing microalgae adherence to the electrode surfaces, and as a

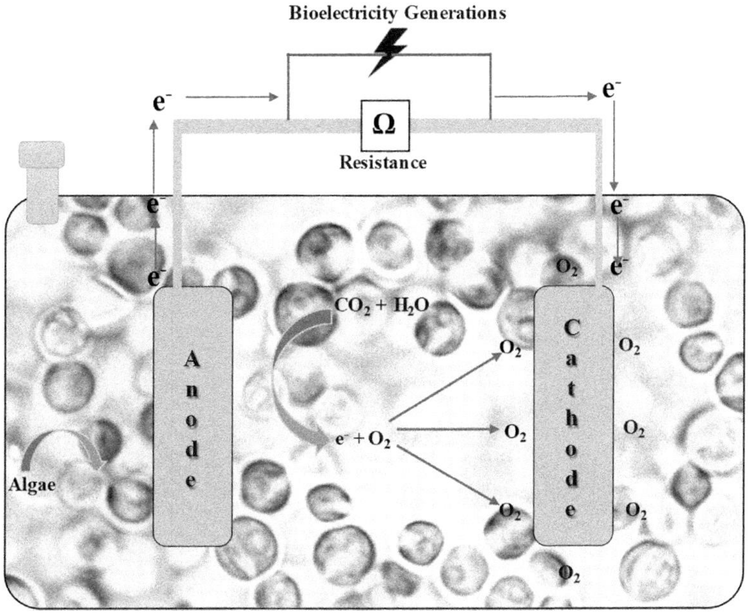

Fig. 1 Single-chamber algae-MFCs are shown schematically. *(Adapted and reproduced from reference R.G. Saratale, C. Kuppam, A. Mudhoo, G.D. Saratale, S. Periyasamy, G. Zhen, ..., G. Kumar, Bioelectrochemical systems using microalgae–a concise research update. Chemosphere, 177 (2017) 35–43.)*

consequence, it may be employed as a photosynthetic biocatalyst for bioelectricity production (Fig. 1). Nishio et al. [78] used photosynthesis-grown *Chlamydomonas reinhardtii* in "single chamber MFC bioreactors" with algae employing *Lactobacillus* and *Geobacter*, as well as found that their framework was capable of yielding a power density of 0.078 W m^2. Yuan et al. [79] investigated a single chamber MFC driven by blue-green algae, and the device revealed a significant chemical oxygen demand and total nitrogen of about 78.9% and 96.8%, respectively.

3.3 Double chamber algae-microbial fuel cell system

A "proton exchange membrane" separates two chambers in this configuration (Figs. 2 and 3). Gajda et al. [80] have shown that an MFC with anodic anaerobic biofilms could produce power whereas the oxidoreduction cycle including algal biomass development in the phototrophic biofilm at the cathodic end produced oxygen. This method produced biomass, cleansed wastewater, and generated power with perfection. Several microalgae species have been employed in an MFC with two compartments, notably *Chlorella vulgaris* [47,66,81], Spirulina platensis [82], and Pseudokirchneriella subcapitata [83]. Scientists found that hybrid algal cultures could also be employed to make MFC cathodes [85]. Strik et al. [84] demonstrated power generation for 100 days in an open system utilizing a mediator-free photosynthetic algal microbial fuel cell, revealing that

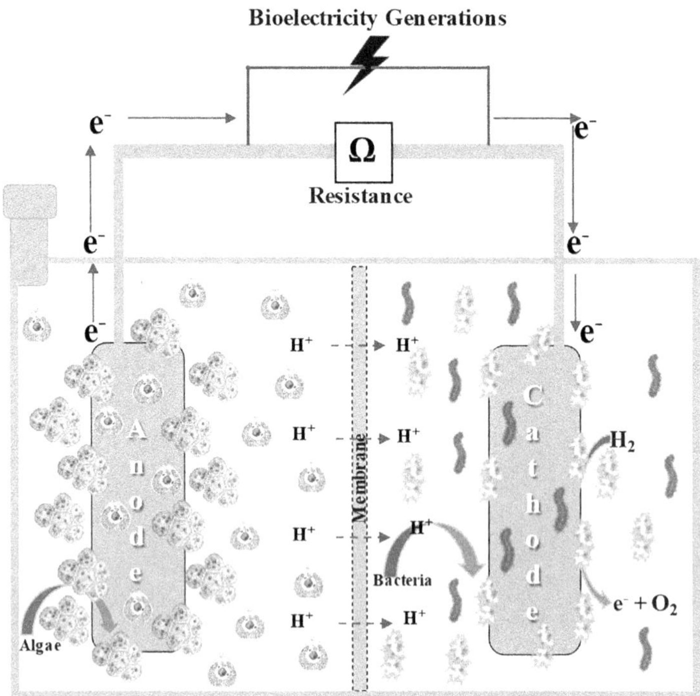

Fig. 2 A double chamber MFC is shown schematically. *(Adapted and reproduced from reference R.G. Saratale, C. Kuppam, A. Mudhoo, G.D. Saratale, S. Periyasamy, G. Zhen, ..., G. Kumar, Bioelectrochemical systems using microalgae–a concise research update. Chemosphere, 177 (2017) 35–43.)*

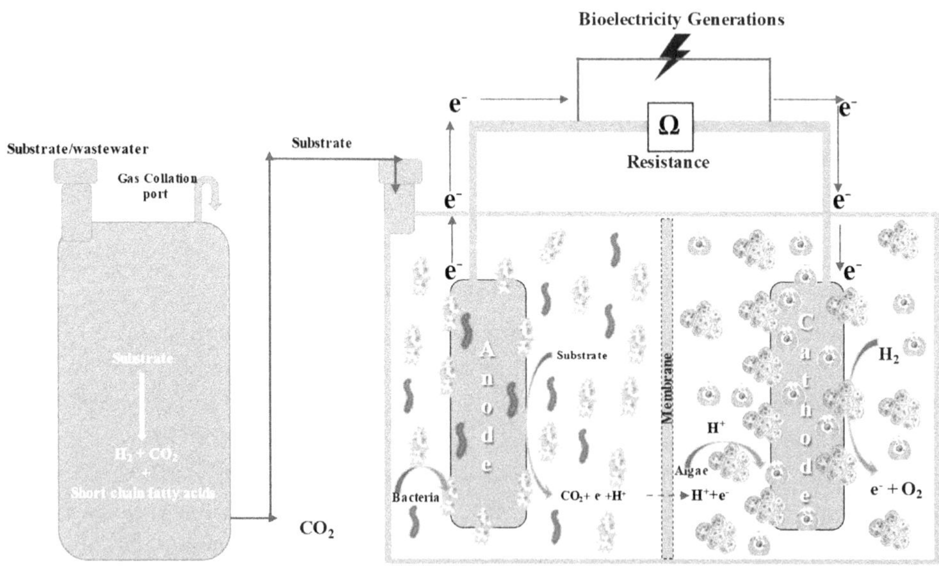

Fig. 3 Diagram of a double chamber MFC with bacteria in the anode and microalgae in the cathode that is combined with a wastewater treatment/bioreactor that produces hydrogen. *(Adapted and reproduce from reference R.G. Saratale, C. Kuppam, A. Mudhoo, G.D. Saratale, S. Periyasamy, G. Zhen, ..., G. Kumar, Bioelectrochemical systems using microalgae–a concise research update. Chemosphere, 177 (2017) 35–43.)*

the MFC's maximum power output capability was 110 mW m^2 of photobioreactor surface area. Mitra and Hill [86] developed an MFC system that ran under continuous flow regimes and employed *Saccharomyces cerevisiae* at the cathodic end and *C. vulgaris* at the anodic end. In this investigation, the peak power density was found to be 0.6 mW m^2. Using green algae in the cathodic chamber at 685 mW m^2 with a 1000 U resistor, Zhang et al. [66] demonstrated excellent nutrient removal and energy generation. As a result of algae growth providing oxygen that limits the system's ability to operate for a longer period, MFC systems tend to operate poorly with little power output for many years to come [66,87].

3.4 Photosynthetic sediment microbial fuel cell systems (PSMFCs)

Photosynthetic-sediment MFC is made up of a cathode compartment on top of the sediment and a cathode compartment within the sediment, both of which include microalgae. The algae cells in the cathode compartment consume the CO_2 created by anodic bacterial activities. Following that, oxygen is produced, and power is generated. He et al. [74] created a "sediment-type MFC" that produced a high voltage of 0.054 0.002 mA at a resistance of 1 kU in a system that had been operational for more than 145 days. Commault et al. [88] recently built a membrane-less sediment-type MFC incorporating a photosynthetic biocathode containing a complex microbial community together with microalgae and cyanobacteria that was capable of producing a maximum power density of 11 mW m^2 for 180 days without feeding.

3.5 Algae-based microbial carbon capture cell (MCC)

Several researchers have shown how effectively an algae-based microbial carbon capture cell (MCC) functioned when lighted. When Liu et al. [50] tested this system in both light and dark settings, they observed that light illumination resulted in a peak power density of 187 mW m^2, which was much higher than the dark-condition output of 21 mW m^2. Liu et al. [50] discovered the role of algal photosynthesis in these systems. Pandit et al. [42] recently investigated Anabaena biocathode performance with MCC utilizing a CO_2 air combination, and they discovered a higher peak power generation compared to the biocathode that had merely been sparged with air. Previously, Wang et al. [47] developed an MCC with a peak power density of 5.6 W m^3, and *C. vulgaris* was employed to decrease CO_2 emissions, with significant CO_2 reductions documented. Researchers discovered that utilizing fixed cells rather than floating cells might increase columbic efficiency by up to 88% [89].

3.6 Anode-catalyzed microalgae microbial fuel cell system

Few studies have used bacteria or microalgae that can produce electrons in the anode compartment and transmit those electrons to the anode without the need for 3.5 Integrated photo-bioelectrochemical systems electro mediators. Live *Chlorella pyrenoidosa* was used by Xu et al. [90] in the anode of an MFC where this species has served as an electron

Table 1 Microalgae-based MFC systems for bioelectricity production.

Reference	Algal used	Power density	Electrodes	
			Anode	Cathode
**[48]	*Scenedesmus obliquus*	153 mW m^{-2}	Plain carbon paper	platinum-coated carbon paper
**[92]	*Chlorella vulgaris*	13.5 mW m^{-2}	Toray carbon cloths	Toray carbon cloths with 10% Teflon
**[81]	*C. vulgaris*	8.79 mW m^{-2}	Carbon felt	Carbon fiber cloth
**[93]	*C. vulgaris*	1926 mW m^{-2}	Carbon fiber brush	Carbon cloth
**[94]	*C. vulgaris*	62.7 mW m^{-2}	Plain graphite	Plain graphite
*[95]	*Escherichia coli*	91 mW m^{-2}	Mn4þgraphite	Fe graphite
**[10]	*Sludge wastewater*	125 mW cm^{-2}	Graphite carbon	Graphite carbon cathode
**[66]	*Geobacterspp*	0.16e1.14 A m^{-2}	Graphite	Graphite
*[66]	*C. vulgaris*	0.068 W m^{-2}	Carbon paper	Carbon paper
**[82]	*Spirulina platensis*	6.5 mW m^{-2}	Platinum electrodes	Platinum cathode
**[96]	*Arthrospira maxima*	20.5 mW m^{-2}	Graphite	Graphite
*[97]	*C. vulgaris*	2.7 mW m^{-2}	Glassy graphite rods	Glassy graphite rods
**[98]	*S. obliquus*	102 mW m^{-2}	Toray carbon paper	Toray carbon paper

Note: MFC design (**Double chamber) and (*Single chamber).
Adapted and reproduced from reference R.G. Saratale, C. Kuppam, A. Mudhoo, G.D. Saratale, S. Periyasamy, G. Zhen, ..., G. Kumar, Bioelectrochemical systems using microalgae–a concise research update. Chemosphere, 177 (2017) 35–43.

donor. A high-power density of 6030 mW m^2 was attained at ideal oxygen levels, algal cell population densities, and incoming light intensities. Additionally, it has been observed that some microbes, like *Rhodopseudomonas* and other purple non-sulfur microbes, may efficiently exploit the biomass contained in an MFC's anode compartment [91]. Table 1 summarizes key findings from several research that examined the performance of anode-catalyzed microalgae.

3.7 Integrated photo-bioelectrochemical system

By integrating an MFC into an algae-based bioreactor, several researchers have created integrated photo-bioelectrochemical systems. An effective method for producing algal biomass and power has been discovered. With a concurrent peak power density yield of 2.2 W m^3, Xiao et al. [83] found considerable removal of COD of up to 92%, ammonium nitrogen removal of 98%, and phosphate removal of 82%. An integrated system was created by Strik et al. [84] by annexing a glass photobioreactor to an MFC for the production of algal biomass and energy. Similarly, Jiang et al. [85] showed that the generation of energy and wastewater cleanup were both possible using an up-flow MFC-photobioreactor linked system. Verstraete et al. [99] have developed a closed-loop mechanism to convert solar energy into biogas. In this work of Verstraete et al. [99], the produced algal biomass was used as feedstock in an anaerobic tank, and under the particular experimental circumstances, a biogas production of $0.5\,N\,m^3$ kg 1 algae and an algal biomass output of 24–30 t VS ha^1 $year^1$ were recorded. There is evidence to indicate the viability of integrated systems for the concurrent production of several forms of biofuels at a relatively low cost and with minimal environmental effect.

3.8 Algae as substrate supplier in dark microbial fuel cell system anodic end

According to the literature, both live and dried algae biomass in the anode compartment of MFCs contain sufficient amounts of carbohydrates, proteins, and lipids for the production of electricity [100]. Pollution reduction and affordable feedstock in MFC processing were two advantages of using microalgae biomass. Because certain microalgae have a very high concentration of cellulose and hemicellulose, pretreatment of the algal biomass is frequently necessary to improve the process efficiency. Also evaluated as a substrate in MFC for the development of oxidizing bacteria at the anodic end is dry algae biomass [93,94,101,102]. *C. vulgaris* and *Ulva lactuca* feedstocks in dry powder have been evaluated by Velasquez-Orta et al. [101] at the anodic end of the MFC system they constructed, they subsequently discovered that *U. lactuca* had a peak power density of $760\,mW\,m^2$ and *C. vulgaris* had a peak power density of $0.98\,W\,m^2$. Rashid et al. [102] employed Scenedesmus algal biomass and activated sludge as a source of nutrients at the anode, they found that increasing the microbial digestibility of the algae by sonication and heat pre-treatment also improved the performance of the MFC test unit.

4 Scope of valuable by-product recovery using microalgae

Photosynthetic microorganisms, particularly microalgae-based biomass, constitute large concentrations of complex carbohydrates, peptides, as well as fats, making them a promising ingredient for antibiotic production, bioenergy manufacturing (e.g., biodiesel, bioethanol, biobutanol, biohydrogen), and many other uses like livestock feed (e.g., pigments, polyunsaturated fatty acids, antioxidants) [103]. Furthermore, some specialized algae species may be cultivated in the cathodic chamber of MCC while simultaneously treating wastewater, producing significant products after harvesting [104]. Cosmetic businesses use additional beneficial chemicals extracted from a variety of different algae species, including *Botryococcus* sp., *Chlorella* sp., *Dunaliella* sp., *Haematococcus* sp., *Phaeodactylum* sp., *Porphyridium* sp., and *Spirulina* sp. [105,106]. Also well-established in the literature is the synthesis of bio-hydrogen from algal biomass using *Scenedesmus obliquus* and

Table 2 List of valuable by-products recovered using microalgae.

S. No.	Microalgae	Valuable by-product	Reference
1	*Dunaliella salina*	Vitamin C supplement and as a food coloring agent for the preparation of orange juices	[108,109]
2	*Fucus vesiculosus* and *Turbinaria conoides*	Antioxidant compounds are used in anti-aging creams to prevent skin disorders	[110]
3	*Botryococcus* sp., *Chlorella* sp., *Dunaliella* sp., *Haematococcus* sp., *Phaeodactylum* sp., *Porphyridium* sp., and *Spirulina* sp.	Cosmetic industries	[105,106]
4	*Scenedesmus obliquus* and *Chlamydomonas reinhardtii*	Bio-hydrogen production employed in the cathodic chamber of MCC	[107]
5	*Haematococcus pluvialis*	Astaxanthin	[111]
6	*Spirulina (Arthrospira),* red algae such as *Porphyridium, Rhodella,* and *Galdieria,* and the *Cryptophyta* and *Glaucophyta*	Food and Cosmetic colors (phycobilin pigments phycocyanin, phycoerythrin, and allophycocyanin)	[112]
7	*Crypthecodinium cohnii*	DHA-rich oil	[113]
8	*zygomycete* fungus *Mortiriella alpina,* although the green alga *Parietochloris incisa*	Arachidonic acid-rich oil	[114]
9	Red algae such as *Porphyridium* and *Rhodella*	Polysaccharides	[115]
10	Cyanobacteria such as *Spirulina (Arthrospira)* and *Synechocystis*	Poly-3-hydroxybutyrate (P3HB)	[116,117]
11	*Dolabella auricularia and Symploca* sp.	Dolastatin 10, monomethyl auristatin E	[118,119]
12	Red macroalga *Porphyra umbilicalis*	Amino acids shionorine and porphyra-334 (Trade name: Helioguard 365) are used in cosmetics	[105]

C. reinhardtii in the cathodic chamber of MCC [107]. Valuable by-products recovered using microalgae are shown in Table 2.

5 Future perspectives

Because of their low yield and inability to withstand performance parameters, the ideas and implementations of BESs-based technologies are still in their infancy [120]. Before modern innovations can be successfully applied to a large extent, their energy efficacy and carbon sequestration must be significantly enhanced. Scaling up BES technology to prepare it for mass production remains a challenge. To establish this, a deep knowledge of electrochemistry, as well as associated biochemistry, must be built, in addition to the creation of effective but economical, cost-effective, and long-term sustainable materials for production. Furthermore, the many parameters influencing the performance of BES must be improved to allow for appropriate scale-up for a green and sustainable field scale deployment of this innovation for capturing CO_2 from diverse origins and transforming it into organo-chemicals and biofuels.

Whether algae-based MFCs technologies can be widely employed for wastewater treatment, more study into the identification of algal strains appropriate for usage in both the anode (substrate) and cathode compartments is required. Following a comprehension and regulation of the systems of the cathodic reaction (electron transfer) and CO_2 uptake to enhance electricity production, bioenergy regulation and recovering of biomass, scale-up as well as modeling of algae-centered MFCs, and finally creation of novel sustainable therapeutic interventions including wastewater treatment centered MFCs or integrated MFCs-photobioreactor strategies, which are primarily used to enhance biomass regeneration and generates value-added products from the algal biomass. Reduced energy production capabilities are one of the primary obstacles in scaling up.

Another significant problem is the development of algal culture efficacy with actual wastewater, since various contaminants and metabolites may inhibit algae growth and/or survival. Pharmaceutical contaminants can even influence and affect the content of activated sludge. As a result, more effective analytical techniques for detecting diverse contaminants at low concentrations are required.

The effectiveness of biodegradation is dependent on the presence of contaminants in the aqueous phase, where algae can eat or interact with them. Some pharmacological compounds are hydrophobic and have low water solubility [121]. A single operating method is insufficient to deal with such contaminants; thus, a multistage system in which hydrophobic and hydrophilic components may be separated and treated based on their solubility is required [122]. The evaluation of a prospective alga can be done in terms of nutrient removal and target molecule breakdown. Immobilization of cells in various matrices is another feature that can help in working with mono and consortium for wastewater treatment. Cell recycling and separation after the procedure are acknowledged as excellent practice [123]. Both natural and artificial matrices may be used for this purpose, but the main disadvantage is limited mass transfer since cells are packed in a border system and interactions with the substrate can be regulated [124]. Furthermore, matrix selection is a difficulty since matrices must not be hazardous to microorganisms or release any toxic byproducts. Algae or consortiums can degrade natural polymers themselves.

Algae and bacteria have different genetic makeups, expression profiles, and metabolic pathways that allow them to degrade pharmaceutical products and other organic pollutants. It is due to differential expression that most wild organisms are not capable of degrading all organic pollutants. A high metabolic burden on cells may arise from genetic improvement techniques that introduce external genes [125]. To assess whether putative microalgal strains are suitable for heterotrophic growth, genome search can be used to screen their sugar transport abilities. Recent metabolic approaches to modify algal genomic DNA for enhanced biomass and biofuel production include genome editing (e.g., CRISPR/Cas9, Zinc-finger nuclease, TALEN), all of which can be incorporated into BES systems for even more efficient biomass and biofuel production [126]. Considering the land and water resources with the greenhouse gas (GHG) emission profiles was necessary for a rigorous life cycle analysis (LCA) of microalgal BESs. A successful establishment of microalgae BESs on a large scale will require national and state subsidies.

Wastewater treatment and environmental conservation can be accomplished simultaneously using algae-based treatment systems. It is important to address challenges to the success of these systems so that they can become commercially sustainable enterprises.

6 Conclusion

Global warming and increased CO_2 concentrations can be addressed with various derivatives of BES, including MCC, photosynthetic microorganisms-microbial fuel cell (P-MFC), and MES. Meanwhile, CO_2 sequestration and wastewater treatment simultaneously can be used to produce valuable products like biofuels, bioenergy, etc. BES-based technologies have yet to be demonstrated at the field scale. A successful field-scale application of these technologies will require optimizing the parameters affecting their yield and performance. In addition, new low-cost electrode materials and membranes will be needed to drastically reduce the cost associated with the operation and fabrication of BESs-based technologies. Multiple small-scale setups can be stacked to circumnavigate the low power output of these technologies. For BES-based technologies to be effective, research must focus on these areas. In addition to their use in conjunction with other treatment technologies such as photodegradation and chemical degradation, algae-bacterial and algae-fungal consortiums can contribute to the cost-effective and efficient treatment of wastewater. Synthetic wastewater systems and systems scaled to laboratories are the focus of most studies available. Field trials with realistically sized samples are needed to evaluate this technology's potential. In contrast to this modern scientific perspective, the evolving challenge of pharmaceutical contaminants will be assisted by heightened community consciousness of non-prescribed drug usage and illegal dumping of expired medication. Policy development to govern pharmaceutical businesses' wastewater management will also aid in resolving the issue.

References

[1] M. Kumar, R. Singh, Performance evaluation of semi continuous vertical flow constructed wetlands (SC-VF-CWs) for municipal wastewater treatment, Bioresour. Technol. 232 (2017) 321–330.

[2] D. Wu, S. Zheng, A. Ding, G. Sun, M. Yang, Performance of a zero valent iron based anaerobic system in swine wastewater treatment, J. Hazard. Mater. 286 (2015) 1–6.

[3] S. Wu, T. Lv, Q. Lu, Z. Ajmal, R. Dong, Treatment of anaerobic digestate supernatant in microbial fuel cell coupled constructed wetlands: evaluation of nitrogen removal, electricity generation, and bacterial community response, Sci. Total Environ. 580 (2017) 339–346.

[4] Y.C. Ho, K.Y. Show, X.X. Guo, I. Norli, F.M. Alkarkhi Abbas, N. Morad, Industrial Discharge and their Effect to the Environment, 2012, Available from: https://www.intechopen.com/.

[5] K. Rabaey, W. Verstraete, Microbial fuel cells: novel biotechnology for energy generation, Trends Biotechnol 23 (6) (2005) 291–298.

[6] H. Luo, G. Liu, R. Zhang, Y. Bai, S. Fu, Y. Hou, Heavy metal recovery combined with H2 production from artificial acid mine drainage using the microbial electrolysis cell, J. Hazard. Mater. 270 (2014) 153–159.

[7] M.L. Sulonen, M.E. Kokko, A.M. Lakaniemi, J.A. Puhakka, Electricity generation from tetrathionate in microbial fuel cells by acidophiles, J. Hazard. Mater. 284 (2015) 182–189.

[8] X. Ju, S. Wu, Y. Zhang, R. Dong, Intensified nitrogen and phosphorus removal in a novel electrolysis-integrated tidal flow constructed wetland system, Water Res. 59 (2014) 37–45.

[9] M. Kumar, R. Singh, Sewage water treatment with energy recovery using constructed wetlands integrated with a bioelectrochemical system, Environ. Sci. Water Res. Technol. 6 (3) (2020) 795–808.

[10] D.R. Bond, D.R. Lovley, Electricity production by Geobacter sulfurreducens attached to electrodes, Appl. Environ. Microbiol. 69 (3) (2003) 1548–1555.

[11] B.H. Kim, H.J. Kim, M.S. Hyun, D.H. Park, Direct electrode reaction of Fe (III)-reducing bacterium, Shewanella putrefaciens, J. Microbiol. Biotechnol. 9 (2) (1999) 127–131.

[12] Z. Li, X. Zhang, L. Lei, Electricity production during the treatment of real electroplating wastewater containing Cr6+ using microbial fuel cell, Process Biochem. 43 (12) (2008) 1352–1358.

[13] Z.D. Liu, H.R. Li, Effects of bio-and abio-factors on electricity production in a mediatorless microbial fuel cell, Biochem. Eng. J. 36 (3) (2007) 209–214.

[14] P. Aelterman, M. Versichele, M. Marzorati, N. Boon, W. Verstraete, Loading rate and external resis tance control the electricity generation of microbial fuel cells with different three-dimensional anodes, Bioresour. Technol. 99 (18) (2008) 8895–8902.

[15] A.P. Borole, C.Y. Hamilton, T. Vishnivetskaya, D. Leak, C. Andras, Improving power production in acetate-fed microbial fuel cells via enrichment of exoelectrogenic organisms in flow-through systems, Biochem. Eng. J. 48 (1) (2009) 71–80.

[16] Z. Liu, J. Liu, S. Zhang, Z. Su, Study of operational performance and electrical response on mediatorless microbial fuel cells fed with carbon-and protein-rich substrates, Biochem. Eng. J. 45 (3) (2009) 185–191.

[17] N. Lu, S.G. Zhou, L. Zhuang, J.T. Zhang, J.R. Ni, Electricity generation from starch processing wastewater using microbial fuel cell technology, Biochem. Eng. J. 43 (3) (2009) 246–251.

[18] S.V. Mohan, G. Mohanakrishna, P.N. Sarma, Effect of anodic metabolic function on bioelectricity generation and substrate degradation in single chambered microbial fuel cell, Environ. Sci. Technol. 42 (21) (2008) 8088–8094.

[19] S.V. Mohan, G. Mohanakrishna, P.N. Sarma, Composite vegetable waste as renewable resource for bioelectricity generation through non-catalyzed open-air cathode microbial fuel cell, Bioresour. Technol. 101 (3) (2010) 970–976.

[20] S.V. Mohan, G. Mohanakrishna, B.P. Reddy, R. Saravanan, P.N. Sarma, Bioelectricity generation from chemical wastewater treatment in mediatorless (anode) microbial fuel cell (MFC) using selectively enriched hydrogen producing mixed culture under acidophilic microenvironment, Biochem. Eng. J. 39 (1) (2008) 121–130.

[21] H. Liu, R. Ramnarayanan, B.E. Logan, Production of electricity during wastewater treatment using a single chamber microbial fuel cell, Environ. Sci. Technol. 38 (7) (2004) 2281–2285.

[22] S. Oh, B.E. Logan, Hydrogen and electricity production from a food processing wastewater using fermentation and microbial fuel cell technologies, Water Res. 39 (19) (2005) 4673–4682.
[23] Y. Feng, X. Wang, B.E. Logan, H. Lee, Brewery wastewater treatment using air-cathode microbial fuel cells, Appl. Microbiol. Biotechnol. 78 (5) (2008) 873–880.
[24] G. Mohanakrishna, S.V. Mohan, P.N. Sarma, Bio-electrochemical treatment of distillery wastewater in microbial fuel cell facilitating decolorization and desalination along with power generation, J. Hazard. Mater. 177 (1–3) (2010) 487–494.
[25] Q. Wen, Y. Wu, D. Cao, L. Zhao, Q. Sun, Electricity generation and modeling of microbial fuel cell from continuous beer brewery wastewater, Bioresour. Technol. 100 (18) (2009) 4171–4175.
[26] A.S. Mathuriya, V.N. Sharma, Bioelectricity production from paper industry waste using a microbial fuel cell by Clostridium species, J. Biochem. Technol. 1 (2) (2009) 49–52.
[27] P. Cheng, R. Shan, H.R. Yuan, W.J. Shen, Y. Chen, Bioelectricity generation from the salinomycin-simulated livestock sewage in a Rhodococcus pyridinivorans inoculated microbial fuel cell, Process Saf. Environ. Prot. 138 (2020) 76–79.
[28] S.A. Patil, V.P. Surakasi, S. Koul, S. Ijmulwar, A. Vivek, Y.S. Shouche, B.P. Kapadnis, Electricity generation using chocolate industry wastewater and its treatment in activated sludge based microbial fuel cell and analysis of developed microbial community in the anode chamber, Bioresour. Technol. 100 (21) (2009) 5132–5139.
[29] S.V. Mohan, S. Srikanth, P.N. Sarma, Non-catalyzed microbial fuel cell (MFC) with open air cathode for bioelectricity generation during acidogenic wastewater treatment, Bioelectrochemistry 75 (2) (2009) 130–135.
[30] Z. Ren, T.E. Ward, J.M. Regan, Electricity production from cellulose in a microbial fuel cell using a defined binary culture, Environ. Sci. Technol. 41 (13) (2007) 4781–4786.
[31] X. Wang, Y. Feng, H. Wang, Y. Qu, Y. Yu, N. Ren, B.E. Logan, Bioaugmentation for electricity generation from corn Stover biomass using microbial fuel cells, Environ. Sci. Technol. 43 (15) (2009) 6088–6093.
[32] J.R. Kim, J. Dec, M.A. Bruns, B.E. Logan, Removal of odors from swine wastewater by using microbial fuel cells, Appl. Environ. Microbiol. 74 (8) (2008) 2540–2543.
[33] K. Ullah, M. Ahmad, V.K. Sharma, P. Lu, A. Harvey, M. Zafar, S. Sultana, Assessing the potential of algal biomass opportunities for bioenergy industry: a review, Fuel 143 (2015) 414–423.
[34] A. Guldhe, P. Singh, N. Renuka, F. Bux, Biodiesel synthesis from wastewater grown microalgal feedstock using enzymatic conversion: a greener approach, Fuel 237 (2019) 1112–1118.
[35] F. Javed, M. Aslam, N. Rashid, Z. Shamair, A.L. Khan, M. Yasin, A.A. Bazmi, Microalgae-based biofuels, resource recovery and wastewater treatment: a pathway towards sustainable biorefinery, Fuel 255 (2019) 115826.
[36] Z. Zhang, D. Sun, T. Wu, Y. Li, Y. Lee, J. Liu, F. Chen, The synergistic energy and carbon metabolism under mixotrophic cultivation reveals the coordination between photosynthesis and aerobic respiration in Chlorella zofingiensis, Algal Res. 25 (2017) 109–116.
[37] M. Nayak, A. Karemore, R. Sen, Performance evaluation of microalgae for concomitant wastewater bioremediation, CO2 biofixation and lipid biosynthesis for biodiesel application, Algal Res. 16 (2016) 216–223.
[38] S.A. Razzak, M. Ilyas, S.A.M. Ali, M.M. Hossain, Effects of CO2 concentration and pH on mixotrophic growth of Nannochloropsis oculata, Appl. Biochem. Biotechnol. 176 (5) (2015) 1290–1302.
[39] S.S. Kumar, V. Kumar, R. Kumar, S.K. Malyan, A. Pugazhendhi, Microbial fuel cells as a sustainable platform technology for bioenergy, biosensing, environmental monitoring, and other low power device applications, Fuel 255 (2019) 115682.
[40] S.S. Kumar, V. Kumar, S.K. Malyan, J. Sharma, T. Mathimani, M.S. Maskarenj, A. Pugazhendhi, Microbial fuel cells (MFCs) for bioelectrochemical treatment of different wastewater streams, Fuel 254 (2019) 115526.
[41] J. Wei, P. Liang, X. Huang, Recent progress in electrodes for microbial fuel cells, Bioresour. Technol. 102 (20) (2011) 9335–9344.
[42] S. Pandit, B.K. Nayak, D. Das, Microbial carbon capture cell using cyanobacteria for simultaneous power generation, carbon dioxide sequestration and wastewater treatment, Bioresour. Technol. 107 (2012) 97–102.

[43] L. Huang, J.M. Regan, X. Quan, Electron transfer mechanisms, new applications, and performance of biocathode microbial fuel cells, Bioresour. Technol. 102 (1) (2011) 316–323.
[44] G. Zhang, Q. Zhao, Y. Jiao, K. Wang, D.J. Lee, N. Ren, Efficient electricity generation from sewage sludge using biocathode microbial fuel cell, Water Res. 46 (1) (2012) 43–52.
[45] S.V. Mohan, S. Srikanth, P. Chiranjeevi, S. Arora, R. Chandra, Algal biocathode for in situ terminal electron acceptor (TEA) production: synergetic association of bacteria–microalgae metabolism for the functioning of biofuel cell, Bioresour. Technol. 166 (2014) 566–574.
[46] X. Hu, B. Liu, J. Zhou, R. Jin, S. Qiao, G. Liu, CO2 fixation, lipid production, and power generation by a novel air-lift-type microbial carbon capture cell system, Environ. Sci. Technol. 49 (17) (2015) 10710–10717.
[47] X. Wang, Y. Feng, J. Liu, H. Lee, C. Li, N. Li, N. Ren, Sequestration of CO2 discharged from anode by algal cathode in microbial carbon capture cells (MCCs), Biosens. Bioelectron. 25 (12) (2010) 2639–2643.
[48] R. Kakarla, B. Min, Photoautotrophic microalgae Scenedesmus obliquus attached on a cathode as oxygen producers for microbial fuel cell (MFC) operation, Int. J. Hydrog. Energy 39 (19) (2014) 10275–10283.
[49] L. Gonzalez Olias, P.J. Cameron, M. Di Lorenzo, Effect of electrode properties on the performance of a photosynthetic microbial fuel cell for atrazine detection, Front. Energy Res. 7 (2019) 105.
[50] T. Liu, L. Rao, Y. Yuan, L. Zhuang, Bioelectricity generation in a microbial fuel cell with a self-sustainable photocathode, Sci. World J. 2015 (2015).
[51] D.B. Wang, T.S. Song, T. Guo, Q. Zeng, J. Xie, Electricity generation from sediment microbial fuel cells with algae-assisted cathodes, Int. J. Hydrog. Energy 39 (25) (2014) 13224–13230.
[52] C.Y. Chen, J.S. Chang, H.Y. Chang, T.Y. Chen, J.H. Wu, W.L. Lee, Enhancing microalgal oil/lipid production from Chlorella sorokiniana CY1 using deep-sea water supplemented cultivation medium, Biochem. Eng. J. 77 (2013) 74–81.
[53] M.H. Osman, A.A. Shah, F.C. Walsh, Recent progress and continuing challenges in bio-fuel cells. Part I: enzymatic cells, Biosens. Bioelectron. 26 (7) (2011) 3087–3102.
[54] M. Rosenbaum, Z. He, L.T. Angenent, Light energy to bioelectricity: photosynthetic microbial fuel cells, Curr. Opin. Biotechnol. 21 (3) (2010) 259–264.
[55] D.P. Strik, R.A. Timmers, M. Helder, K.J. Steinbusch, H.V. Hamelers, C.J. Buisman, Microbial solar cells: applying photosynthetic and electrochemically active organisms, Trends Biotechnol. 29 (1) (2011) 41–49.
[56] X. Cao, X. Huang, P. Liang, K. Xiao, Y. Zhou, X. Zhang, B.E. Logan, A new method for water desalination using microbial desalination cells, Environ. Sci. Technol. 43 (18) (2009) 7148–7152.
[57] K.S. Jacobson, D.M. Drew, Z. He, Efficient salt removal in a continuously operated upflow microbial desalination cell with an air cathode, Bioresour. Technol. 102 (1) (2011) 376–380.
[58] M. Mehanna, P.D. Kiely, D.F. Call, B.E. Logan, Microbial electrodialysis cell for simultaneous water desalination and hydrogen gas production, Environ. Sci. Technol. 44 (24) (2010) 9578–9583.
[59] D. Pant, A. Singh, G. Van Bogaert, S.I. Olsen, P.S. Nigam, L. Diels, K. Vanbroekhoven, Bioelectrochemical systems (BES) for sustainable energy production and product recovery from organic wastes and industrial wastewaters, RSC Adv. 2 (4) (2012) 1248–1263.
[60] M. Winter, R.J. Brodd, What are batteries, fuel cells, and supercapacitors? Chem. Rev. 104 (10) (2004) 4245–4270.
[61] R.H. Mahmoud, H. Abd El-Raheem, R.Y. Hassan, Algae-mediated bioelectrochemical system: The future of algae in the electrochemical operations, in: Bio-Electrochemical Systems, CRC Press, 2022, pp. 93–123.
[62] F. Delrue, P.D. Álvarez-Díaz, S. Fon-Sing, G. Fleury, J.F. Sassi, The environmental biorefinery: using microalgae to remediate wastewater, a win-win paradigm, Energies 9 (3) (2016) 132.
[63] R. Huarachi-Olivera, A. Dueñas-Gonza, U. Yapo-Pari, P. Vega, M. Romero-Ugarte, J. Tapia, M. Esparza, Bioelectrogenesis with microbial fuel cells (MFCs) using the microalga Chlorella vulgaris and bacterial communities, Electron. J. Biotechnol. 31 (2018) 34–43.
[64] A. Colombo, S. Marzorati, G. Lucchini, P. Cristiani, D. Pant, A. Schievano, Assisting cultivation of photosynthetic microorganisms by microbial fuel cells to enhance nutrients recovery from wastewater, Bioresour. Technol. 237 (2017) 240–248.

[65] H. Li, Y. Tian, W. Zuo, J. Zhang, X. Pan, L. Li, X. Su, Electricity generation from food wastes and characteristics of organic matters in microbial fuel cell, Bioresour. Technol. 205 (2016) 104–110.
[66] Y. Zhang, J.S. Noori, I. Angelidaki, Simultaneous organic carbon, nutrients removal and energy production in a photomicrobial fuel cell (PFC), Energy Environ. Sci. 4 (10) (2011) 4340–4346.
[67] Q. Hou, C. Nie, H. Pei, W. Hu, L. Jiang, Z. Yang, The effect of algae species on the bioelectricity and biodiesel generation through open-air cathode microbial fuel cell with kitchen waste anaerobically digested effluent as substrate, Bioresour. Technol. 218 (2016) 902–908.
[68] A.E. Inglesby, A.C. Fisher, Enhanced methane yields from anaerobic digestion of Arthrospira maxima biomass in an advanced flow-through reactor with an integrated recirculation loop microbial fuel cell, Energy Environ. Sci. 5 (7) (2012) 7996–8006.
[69] A.S. Mathuriya, J.V. Yakhmi, Microbial fuel cells—applications for generation of electrical power and beyond, Crit. Rev. Microbiol. 42 (1) (2016) 127–143.
[70] B.E. Logan, J.M. Regan, Microbial fuel cells—challenges and applications, Environ. Sci. Technol. 40 (17) (2006) 5172–5180.
[71] S. Bajracharya, K. Vanbroekhoven, C.J. Buisman, D. Pant, D.P. Strik, Application of gas diffusion biocathode in microbial electrosynthesis from carbon dioxide, Environ. Sci. Pollut. Res. 23 (2016) 22292–22308.
[72] K. Wang, W. Li, X. Gong, Y. Li, C. Wu, N. Ren, Spectral study of dissolved organic matter in biosolid during the composting process using inorganic bulking agent: UV–vis, GPC, FTIR and EEM, Int. Biodeterior. Biodegrad. 85 (2013) 617–623.
[73] E.E. Powell, G.A. Hill, Economic assessment of an integrated bioethanol–biodiesel–microbial fuel cell facility utilizing yeast and photosynthetic algae, Chem. Eng. Res. Des. 87 (9) (2009) 1340–1348.
[74] Z. He, J. Kan, F. Mansfeld, L.T. Angenent, K.H. Nealson, Self-sustained phototrophic microbial fuel cells based on the synergistic cooperation between photosynthetic microorganisms and heterotrophic bacteria, Environ. Sci. Technol. 43 (5) (2009) 1648–1654.
[75] W. Logroño, M. Pérez, G. Urquizo, A. Kadier, M. Echeverría, C. Recalde, G. Rákhely, Single chamber microbial fuel cell (SCMFC) with a cathodic microalgal biofilm: a preliminary assessment of the generation of bioelectricity and biodegradation of real dye textile wastewater, Chemosphere 176 (2017) 378–388.
[76] C.C. Lin, C.H. Wei, C.I. Chen, C.J. Shieh, Y.C. Liu, Characteristics of the photosynthesis microbial fuel cell with a Spirulina platensis biofilm, Bioresour. Technol. 135 (2013) 640–643.
[77] C.C. Fu, C.H. Su, T.C. Hung, C.H. Hsieh, D. Suryani, W.T. Wu, Effects of biomass weight and light intensity on the performance of photosynthetic microbial fuel cells with Spirulina platensis, Bioresour. Technol. 100 (18) (2009) 4183–4186.
[78] K. Nishio, K. Hashimoto, K. Watanabe, Light/electricity conversion by defined cocultures of Chlamydomonas and Geobacter, J. Biosci. Bioeng. 115 (4) (2013) 412–417.
[79] Y. Yuan, B. Zhao, S. Zhou, S. Zhong, L. Zhuang, Electrocatalytic activity of anodic biofilm responses to pH changes in microbial fuel cells, Bioresour. Technol. 102 (13) (2011) 6887–6891.
[80] I. Gajda, J. Greenman, C. Melhuish, I. Ieropoulos, Self-sustainable electricity production from algae grown in a microbial fuel cell system, Biomass Bioenergy 82 (2015) 87–93.
[81] G.J. Zhou, F.Q. Peng, L.J. Zhang, G.G. Ying, Biosorption of zinc and copper from aqueous solutions by two freshwater green microalgae Chlorella pyrenoidosa and Scenedesmus obliquus, Environ. Sci. Pollut. Res. 19 (2012) 2918–2929.
[82] C.C. Fu, T.C. Hung, W.T. Wu, T.C. Wen, C.H. Su, Current and voltage responses in instant photosynthetic microbial cells with Spirulina platensis, Biochem. Eng. J. 52 (2–3) (2010) 175–180.
[83] L. Xiao, E.B. Young, J.A. Berges, Z. He, Integrated photo-bioelectrochemical system for contaminants removal and bioenergy production, Environ. Sci. Technol. 46 (20) (2012) 11459–11466.
[84] D.P. Strik, H. Terlouw, H.V. Hamelers, C.J. Buisman, Renewable sustainable biocatalyzed electricity production in a photosynthetic algal microbial fuel cell (PAMFC), Appl. Microbiol. Biotechnol. 81 (2008) 659–668.
[85] Y. Jiang, T. Yoshida, A. Quigg, Photosynthetic performance, lipid production and biomass composition in response to nitrogen limitation in marine microalgae, Plant Physiol. Biochem. 54 (2012) 70–77.

[86] P. Mitra, G.A. Hill, Continuous microbial fuel cell using a photoautotrophic cathode and a fermentative anode, Can. J. Chem. Eng. 90 (4) (2012) 1006–1010.

[87] A.J. McCormick, P. Bombelli, A.M. Scott, A.J. Philips, A.G. Smith, A.C. Fisher, C.J. Howe, Photosynthetic biofilms in pure culture harness solar energy in a mediatorless bio-photovoltaic cell (BPV) system, Energy Environ. Sci. 4 (11) (2011) 4699–4709.

[88] A.S. Commault, G. Lear, R.J. Weld, Maintenance of geobacter-dominated biofilms in microbial fuel cells treating synthetic wastewater, Bioelectrochemistry 106 (2015) 150–158.

[89] M. Zhou, M. Chi, H. Wang, T. Jin, Anode modification by electrochemical oxidation: a new practical method to improve the performance of microbial fuel cells, Biochem. Eng. J. 60 (2012) 151–155.

[90] C. Xu, K. Poon, M.M. Choi, R. Wang, Using live algae at the anode of a microbial fuel cell to generate electricity, Environ. Sci. Pollut. Res. 22 (2015) 15621–15635.

[91] D. Xing, S. Cheng, J.M. Regan, B.E. Logan, Change in microbial communities in acetate-and glucose-fed microbial fuel cells in the presence of light, Biosens. Bioelectron. 25 (1) (2009) 105–111.

[92] A.G. Del Campo, P. Cañizares, M.A. Rodrigo, F.J. Fernández, J. Lobato, Microbial fuel cell with an algae-assisted cathode: a preliminary assessment, J. Power Sources 242 (2013) 638–645.

[93] Y. Cui, N. Rashid, N. Hu, M.S.U. Rehman, J.I. Han, Electricity generation and microalgae cultivation in microbial fuel cell using microalgae-enriched anode and bio-cathode, Energy Convers. Manag. 79 (2014) 674–680.

[94] L. Gouveia, C. Neves, D. Sebastião, B.P. Nobre, C.T. Matos, Effect of light on the production of bioelectricity and added-value microalgae biomass in a photosynthetic alga microbial fuel cell, Bioresour. Technol. 154 (2014) 171–177.

[95] D.H. Park, J.G. Zeikus, Improved fuel cell and electrode designs for producing electricity from microbial degradation, Biotechnol. Bioeng. 81 (3) (2003) 348–355.

[96] A.E. Inglesby, K. Yunus, A.C. Fisher, In situ fluorescence and electrochemical monitoring of a photosynthetic microbial fuel cell, Phys. Chem. Chem. Phys. 15 (18) (2013) 6903–6911.

[97] E.E. Powell, R.W. Evitts, G.A. Hill, J.C. Bolster, A microbial fuel cell with a photosynthetic microalgae cathodic half cell coupled to a yeast anodic half cell, Energy Sources A: Recovery Util. Environ. Eff. 33 (5) (2011) 440–448.

[98] S. Kondaveeti, K.S. Choi, R. Kakarla, B. Min, Microalgae Scenedesmus obliquus as renewable biomass feedstock for electricity generation in microbial fuel cells (MFCs), Front. Environ. Sci. Eng. 8 (2014) 784–791.

[99] W. Verstraete, P. Van de Caveye, V. Diamantis, Maximum use of resources present in domestic "used water, Bioresour. Technol. 100 (23) (2009) 5537–5545.

[100] C.H. Liu, D.H. Huang, Reduction of power consumption and carbon footprints by applying multi-objective optimisation via genetic algorithms, Int. J. Prod. Res. 52 (2) (2014) 337–352.

[101] S.B. Velasquez-Orta, T.P. Curtis, B.E. Logan, Energy from algae using microbial fuel cells, Biotechnol. Bioeng. 103 (6) (2009) 1068–1076.

[102] N. Rashid, Y.F. Cui, M.S.U. Rehman, J.I. Han, Enhanced electricity generation by using algae biomass and activated sludge in microbial fuel cell, Sci. Total Environ. 456 (2013) 91–94.

[103] V.C. Akubude, K.N. Nwaigwe, E. Dintwa, Production of biodiesel from microalgae via nanocatalyzed transesterification process: a review, Mater. Sci. Energy Technol. 2 (2) (2019) 216–225.

[104] S. Das, S. Das, I. Das, M.M. Ghangrekar, Application of bioelectrochemical systems for carbon dioxide sequestration and concomitant valuable recovery: a review, Mater. Sci. Energy Technol. 2 (3) (2019) 687–696.

[105] M.A. Borowitzka, High-value products from microalgae—their development and commercialisation, J. Appl. Phycol. 25 (3) (2013) 743–756.

[106] J.N. Rosenberg, G.A. Oyler, L. Wilkinson, M.J. Betenbaugh, A green light for engineered algae: redirecting metabolism to fuel a biotechnology revolution, Curr. Opin. Biotechnol. 19 (5) (2008) 430–436.

[107] S. Kosourov, A. Tsygankov, M. Seibert, M.L. Ghirardi, Sustained hydrogen photoproduction by Chlamydomonas reinhardtii: effects of culture parameters, Biotechnol. Bioeng. 78 (7) (2002) 731–740.

[108] E.W. Becker, Microalgae for human and animal nutrition, in: Handbook of Microalgal Culture: Applied Phycology and Biotechnology, 2013, pp. 461–503.
[109] O. Pulz, W. Gross, Valuable products from biotechnology of microalgae, Appl. Microbiol. Biotechnol. 65 (6) (2004) 635–648.
[110] J.Y. Je, P.J. Park, E.K. Kim, J.S. Park, H.D. Yoon, K.R. Kim, C.B. Ahn, Antioxidant activity of enzymatic extracts from the brown seaweed Undaria pinnatifida by electron spin resonance spectroscopy, LWT-Food Sci. Technol. 42 (4) (2009) 874–878.
[111] G.R. Cysewski, R.T. Lorenz, Industrial production of microalgal cell-mass and secondary products-species of high potential, in: Handbook of Microalgal Culture: Biotechnology and Applied Phycology, 2004, pp. 281–288.
[112] N.T. Eriksen, Production of phycocyanin—a pigment with applications in biology, biotechnology, foods and medicine, Appl. Microbiol. Biotechnol. 80 (1) (2008) 1–14.
[113] J. Wynn, P. Behrens, A. Sundararajan, J. Hansen, K. Apt, Production of single cell oils by dinoflagellates, in: Single cell oils, 79AOCS press, 2010, pp. 115–129.
[114] C. Bigogno, I. Khozin-Goldberg, S. Boussiba, A. Vonshak, Z. Cohen, Lipid and fatty acid composition of the green oleaginous alga Parietochloris incisa, the richest plant source of arachidonic acid, Phytochemistry 60 (5) (2002) 497–503.
[115] S.M. Arad, O. Levy-Ontman, Red microalgal cell-wall polysaccharides: biotechnological aspects, Curr. Opin. Biotechnol. 21 (3) (2010) 358–364.
[116] M. Vincenzini, C. Sili, R. De Philippis, A. Ena, R. Materassi, Occurrence of poly-beta-hydroxybutyrate in Spirulina species, J. Bacteriol. 172 (5) (1990) 2791–2792.
[117] P.S. Toh, M.H. Jau, S.P. Yew, R.M. Abed, K. Sudesh, Comparison of Polyhydroxyalkanoates biosynthesis, obilization and the effects on cellular morphology in Spirulina platensis and Synechocystis sp. Uniwg, J. Biosci. 19 (2) (2008) 21–38.
[118] G.R. Pettit, Y. Kamano, C.L. Herald, A.A. Tuinman, F.E. Boettner, H. Kizu, R.J. Bontems, The isolation and structure of a remarkable marine animal antineoplastic constituent: dolastatin 10, J. Am. Chem. Soc. 109 (22) (1987) 6883–6885.
[119] H. Luesch, R.E. Moore, V.J. Paul, S.L. Mooberry, T.H. Corbett, Isolation of dolastatin 10 from the marine cyanobacterium Symploca species VP642 and total stereochemistry and biological evaluation of its analogue symplostatin 1, J. Nat. Prod. 64 (7) (2001) 907–910.
[120] B.E. Logan, Scaling up microbial fuel cells and other bioelectrochemical systems, Appl. Microbiol. Biotechnol. 85 (6) (2010) 1665–1671.
[121] S. Pilli, B. Sellamuthu, A.K. Pandey, R.D. Tyagi, Treatment of wastewater containing pharmaceuticals: biological treatment, in: Current Developments in Biotechnology and Bioengineering, 2020, pp. 463–520.
[122] S. Choudhary, L. Gupta, S. Rani, K. Dave, U. Gupta, Impact of dendrimers on solubility of hydrophobic drug molecules, Front. Pharmacol. 8 (2017) 261.
[123] M. Kube, B. Jefferson, L. Fan, F. Roddick, The impact of wastewater characteristics, algal species selection and immobilisation on simultaneous nitrogen and phosphorus removal, Algal Res. 31 (2018) 478–488.
[124] E. Eroglu, S.M. Smith, C.L. Raston, Application of various immobilization techniques for algal bioprocesses, in: Biomass and Biofuels from Microalgae, Springer, Cham, 2015, pp. 19–44.
[125] K. Jawed, S.S. Yazdani, M.A. Koffas, Advances in the development and application of microbial consortia for metabolic engineering, Metab. Eng. Commun. 9 (2019) e00095.
[126] Y. Maeda, T. Yoshino, T. Matsunaga, M. Matsumoto, T. Tanaka, Marine microalgae for production of biofuels and chemicals, Curr. Opin. Biotechnol. 50 (2018) 111–120.

Index

Note: Page numbers followed by *f* indicate figures and *t* indicate tables.

B

C

N

O

P

Y

Z

9780323910231